Shale Gas Production Processes

Shale Gas Production Processes

James G. Speight, PhD, DSc
CD&W Inc., Laramie, Wyoming, USA

AMSTERDAM • BOSTON • HEIDELBERG • LONDON
NEW YORK • OXFORD • PARIS • SAN DIEGO
SAN FRANCISCO • SINGAPORE • SYDNEY • TOKYO
Gulf Professional Publishing is an imprint of Elsevier

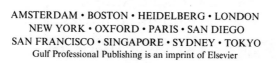

Gulf Professional Publishing is an imprint of Elsevier
The Boulevard, Langford Lane, Kidlington, Oxford, OX5 1GB, UK
225 Wyman Street, Waltham, MA 02451, USA

First published 2013

British Library Cataloguing in Publication Data
A catalogue record for this book is available from the British Library

Library of Congress Cataloging-in-Publication Data
A catalog record for this book is available from the Library of Congress

ISBN: 978-0-12-404571-2

For information on all Gulf Professional Publishing publications
visit our website at store.elsevier.com

This book has been manufactured using Print On Demand technology. Each copy is produced
to order and is limited to black ink. The online version of this book will show color figures
where appropriate.

Transferred to Digital Printing in 2013

Working together
to grow libraries in
developing countries

www.elsevier.com • www.bookaid.org

CONTENTS

Natural gas production from shale formations (*shale gas*) is one of the most rapidly expanding trends in current gas exploration and production. In some cases, this has included bringing drilling and production to regions of the United States that have seen little or no activity in the past. Thus, shale gas has not only changed the energy distribution in the US as a result of this newfound popularity, but shale gas development is also bringing change to the environmental and socioeconomic landscape, particularly in those areas where gas development is new. With these changes have come questions about the nature of shale gas development, the potential environmental impacts, and the ability of the current regulatory structure to deal with these issues.

Shale gas is considered to be unconventional source as the gas may be attached to or adsorbed onto organic matter. The gas is contained in difficult-to-produce reservoirs—shale is rock that can hold huge amounts of gas, not only in the zones between the particles; it must be remembered that some of the particles are organic and can also hold gas like sponges. Evaluation of the shale gas potential of sedimentary basins has now become an important area of development internationally and is of great national interest as shale gas potential evaluation will have a direct and positive impact on the energy security of many countries which have sizable resources in sedimentary basins. However, it will be appreciated that the reserves estimations are not static and are changing annually based upon new discoveries and improvements in drilling and recovery techniques.

The increasing significance of shale gas globally has led to the need for a deeper understanding of shale behavior. Increased understanding of gas shale reservoirs will enable better decision-making regarding the development of these resources. To find these reserves may be easy but the technology to produce gas therefrom is very expensive. The technique involving drilling straight through gas bearing rock meant that the resultant hole had very little exposure to the rock for the purpose of allowing gas to escape. Hydraulic fracturing is the only way to increase such exposure for ensuring a successful gas production rate.

Two decades ago shale gas was of limited importance but, due to issues related to the price and availability of gas at times of natural disasters, concerns grew that natural gas prices would continue to escalate. Thus, the objective of this book is to present an introduction to shale gas resources as well as offer an understanding of the geomechanical properties of shale, the need for hydraulic fracturing, and an indication of shale gas processing. The book also introduces the reader to issues regarding the nature of shale gas development, the potential environmental impacts, and the ability of the current regulatory structure to deal with these issues. The book also serves to introduce scientists, engineers, managers, regulators, and policy makers to objective sources of information upon which to make decisions about meeting and managing the challenges that may arise.

Dr. James G. Speight
Laramie, WY
May 15, 2013

CHAPTER 1

Origin of Shale Gas

1.1 INTRODUCTION

The generic term *natural gas* applies to gases commonly associated with petroliferous (petroleum producing, petroleum containing) geologic formations. Natural gas generally contains high proportions of methane (a single carbon hydrocarbon compound, CH_4) and some of the higher molecular weight higher paraffins (C_nH_{2n+2}) generally containing up to six carbon atoms may also be present in small quantities (Table 1.1). The hydrocarbon constituents of natural gas are combustible, but nonflammable nonhydrocarbon components such as carbon dioxide, nitrogen, and helium are often present in the minority and are regarded as contaminants.

In addition to the natural gas found in petroleum reservoirs, there are also those reservoirs in which natural gas may be the sole occupant. The principal constituent of natural gas is methane, but other hydrocarbons, such as ethane, propane, and butane, may also be present. Carbon dioxide is also a common constituent of natural gas. Trace amounts of rare gases, such as helium, may also occur, and certain natural gas reservoirs are a source of these rare gases. Just as petroleum can vary in composition, so can natural gas. Differences in natural gas composition occur between different reservoirs, and two wells in the same field may also yield gaseous products that are different in composition (Mokhatab et al., 2006; Speight, 2007, 2014).

Natural gas resources are typically divided into two categories: conventional and unconventional (Mokhatab et al., 2006; Speight, 2007, 2014). Conventional gas typically is found in reservoirs with a permeability greater than 1 millidarcy (>1 mD) and can be extracted via traditional techniques. A large proportion of the gas produced globally to date is conventional and is relatively easy and inexpensive to extract. In contrast, unconventional gas is found in reservoirs with relatively low permeability (<1 mD) and hence cannot be extracted by conventional methods.

Table 1.1 Constituents of Natural Gas		
Name	Formula	Vol. %
Methane	CH_4	>85
Ethane	C_2H_6	3–8
Propane	C_3H_8	1–5
Butane	C_4H_{10}	1–2
Pentane[1]	C_5H_{12}	1–5
Carbon dioxide	CO_2	1–2
Hydrogen sulfide	H_2S	1–2
Nitrogen	N_2	1–5
Helium	He	<0.5
Pentane[1]: Pentane and higher molecular weight hydrocarbons, including benzene and toluene (Speight, 2014).		

There are several general definitions that have been applied to natural gas. Thus, *lean* gas is gas in which methane is the major constituent. *Wet* gas contains considerable amounts of the higher molecular weight hydrocarbons. *Sour* gas contains hydrogen sulfide whereas *sweet* gas contains very little, if any, hydrogen sulfide. *Residue* gas is natural gas from which the higher molecular weight hydrocarbons have been extracted and *casing head* gas is derived from petroleum but is separated at the separation facility at the well head.

To further define the terms *dry* and *wet* in quantitative measures, the term *dry* natural gas indicates that there is less than 0.1 gallon (1 gallon, US, = 264.2 m^3) of gasoline vapor (higher molecular weight paraffins) per 1000 ft^3 (1 ft^3 = 0.028 m^3). The term *wet natural gas* indicates that there are such paraffins present in the gas, in fact more than 0.1 gal/1000 ft^3.

1.2 SHALE

Shale formations and silt formations are the most abundant sedimentary rocks in the Earth's crust. In petroleum geology, organic shale formations are source rocks as well as seal rocks that trap oil and gas (Speight, 2014). In reservoir engineering, shale formations are flow barriers. In drilling, the bit often encounters greater shale volumes than reservoir sands. In seismic exploration, shale formations interfacing with other rocks often form good seismic reflectors. As a result, seismic and petrophysical properties of shale formations and the relationships among these properties are important for both exploration and reservoir management. Shale formations are a worldwide occurrence (see Chapter 2).

Shale is a geological rock formation rich in clay, typically derived from fine sediments, deposited in fairly quiet environments at the bottom of seas or lakes, having then been buried over the course of millions of years. Shale formations can serve as pressure barriers in basins, as top seals, and as reservoirs in shale gas plays.

More technically, shale is a fissile, terrigenous sedimentary rock in which particles are mostly of silt and clay size (Blatt and Tracy, 2000). In this definition, *fissile* refers to the ability of the shale to split into thin sheets along the bedding and *terrigenous* refers to the origin of the sediment. In many basins, the fluid pressure of the aqueous system becomes significantly elevated, leading to the formation of a hydrofracture, and fluid bleed-off. However, the occurrence of a natural hydrofracture is an unlikely process in the circumstances that exist in most basins.

When a significant amount of organic matter has been deposited with the sediments, the shale rock can contain organic solid material (kerogen). The properties and composition of shale place it in the category of sedimentary rocks known as *mudstones*. Shale is distinguished from other mudstones because it is laminated and fissile—the shale is composed of many thin layers and readily splits into thin pieces along the laminations.

Shale is composed mainly of clay-size mineral grains, which are usually clay minerals such as illite, kaolinite, and smectite. Shale usually contains other clay-size mineral particles such as quartz, chert, and feldspar. Other constituents might include organic particles, carbonate minerals, iron oxide minerals, sulfide minerals, and heavy mineral grains and the presence of such minerals in shale is determined by the environment under which the shale constituents were.

Shale comes in two general varieties based on organic content: (i) dark or (ii) light. Dark colored or black shale formations are organic rich, whereas the lighter colored shale formations are organic lean. Organic-rich shale formations were deposited under conditions of little or no oxygen in the water, which preserved the organic material from decay. The organic matter was mostly plant debris that had accumulated with the sediment.

Black organic shale formations are the source rock for many of the oil and natural gas deposits of the world. These black shale formations obtain their black color from tiny particles of organic matter that were deposited with the mud from which the shale formed. As the mud was buried and warmed within the earth some of the organic material was transformed into oil and natural gas.

A black color in sedimentary rocks almost always indicates the presence of organic materials. Just 1% or 2% of organic materials can impart a dark gray or black color to the rock. In addition, this black color almost always implies that the shale formed from sediment deposited in an oxygen-deficient environment. Any oxygen that entered the environment quickly reacted with the decaying organic debris. If a large amount of oxygen was present, the organic debris would all have decayed. An oxygen-poor environment also provides the proper conditions for the formation of sulfide minerals such as pyrite, another important mineral found in most black shale sediments or formations.

The presence of organic debris in black shale formations makes them the candidates for oil and gas generation. If the organic material is preserved and properly heated after burial, oil and natural gas might be produced. The Barnett shale, Marcellus shale, Haynesville shale, Fayetteville shale, and other gas producing rocks are all dark gray or black shale formations that yield natural gas.

The oil and natural gas migrated out of the shale and upward through the sediment mass because of their low density. The oil and gas were often trapped within the pore spaces of an overlying rock unit such as a sandstone formation. These types of oil and gas deposits are known as *conventional reservoirs* because the fluids can easily flow through the pores of the rock and into the extraction well.

Shale formations are ubiquitous in sedimentary basins: they typically form about 80% of what a well will drill through. As a result, the main organic-rich shale formations have already been identified in most regions of the world. Their depths vary from near surface to several thousand feet underground, while their thickness varies from tens of feet to several hundred feet. Often, enough is known about the geological history (Table 1.2) to infer which shale formations are likely to contain gas (or oil, or a mixture of both). In that sense there may

Table 1.2 The Geologic Timescale				
Era	Period	Epoch	Approximate Duration (Millions of Years)	Approximate Number of Years Ago (Millions of Years)
Cenozoic	Quaternary	Holocene	10,000 years ago to the present	
		Pleistocene	2	0.01
	Tertiary	Pliocene	11	2
		Miocene	12	13
		Oligocene	11	25
		Eocene	22	36
		Paleocene	71	58
Mesozoic	Cretaceous		71	65
	Jurassic		54	136
	Triassic		35	190
Paleozoic	Permian		55	225
	Carboniferous		65	280
	Devonian		60	345
	Silurian		20	405
	Ordovician		75	425
	Cambrian		100	500
Pre-Cambrian			3380	600

appear to be no real need for a major exploration effort and expense required for shale gas. However, the amount of gas present and particularly the amount of gas that can be recovered technically and economically cannot be known until a number of wells have been drilled and tested.

Each shale formation has different geological characteristics that affect the way gas can be produced, the technologies needed, and the economics of production. Different parts of the (generally large) shale deposits will also have different characteristics: *small sweet spots* or *core areas* may provide much better production than the remainder of the formation, often because of the presence of natural fractures that enhance permeability (Hunter and Young, 1953).

The amount of natural gas liquids (NGLs—hydrocarbons having a higher molecular weight than methane, such as propane, butane, pentane, hexane, heptane, and even octane) commonly associated with natural gas production present in the gas can also vary considerably, with important implications for the economics of production. While most dry gas plays in the United States are probably uneconomic at the current low natural gas prices, plays with significant liquid content can be produced for the value of the liquids only (the market value of NGLs is correlated with oil prices, rather than gas prices), making gas an essentially free by-product.

In the late 1990s, natural gas drilling companies developed new methods for liberating oil and natural gas that is trapped within the tiny pore spaces of shale. This discovery was significant because it unlocked some of the largest natural gas deposits in the world.

The Barnett shale of Texas was the first major natural gas field developed in a shale reservoir rock. Producing gas from the Barnett shale was a challenge because the pore spaces in shale are so tiny that the gas has difficulty moving through the shale and into the well. Drillers discovered that the permeability of the shale could be increased by pumping water down the well under pressure that was high enough to fracture the shale. These fractures liberated some of the gas from the pore spaces and allowed that gas to flow to the well (hydraulic fracturing, hydrofracking).

Horizontal drilling and hydraulic fracturing revolutionized drilling technology and paved the way for developing several giant natural gas fields. These include the Marcellus shale in the Appalachians, the Haynesville shale in Louisiana, and the Fayetteville shale in Arkansas. These enormous shale reservoirs hold enough natural gas to serve all of the United States' needs for 20 years or more.

Hydraulic properties are characteristics of a rock such as permeability and porosity that reflect its ability to hold and transmit fluids such as water, oil, or natural gas. In this respect, shale has a very small particle size so the interstitial spaces are very small. In fact they are so small that oil, natural gas, and water have difficulty moving through the rock. Shale can therefore serve as a cap rock for oil and natural gas traps and it also is an aquiclude that blocks or limits the flow of underground water.

Although the interstitial spaces in a shale formation are very small they can take up a significant volume of the rock. This allows the shale to hold significant amounts of water, gas, or oil but not be able to effectively transmit them because of the low permeability. The oil and gas industry overcomes these limitations of shale by using horizontal drilling and hydraulic fracturing to create artificial porosity and permeability within the rock.

Some of the clay minerals that occur in shale have the ability to absorb or adsorb large amounts of water, natural gas, ions, or other substances. This property of shale can enable it to selectively and tenaciously hold or freely release fluids or ions.

Thus, this shale gas resource can be considered a technology-driven resource as achieving gas production out of otherwise unproductive rock requires technology-intensive processes. Maximizing gas recovery requires far more wells than would be the case in conventional natural gas operations. Furthermore, horizontal wells with horizontal legs up to one mile or more in length are widely used to access the reservoir to the greatest extent possible.

Multistage hydraulic fracturing (see Chapter 3), where the shale is cracked under high pressures at several places along the horizontal section of the well, is used to create conduits through which gas can flow. Microseismic imaging allows operators to visualize where this fracture growth is occurring in the reservoir. However, as a technology-driven resource, the rate of development of shale gas may become limited by the availability of required resources, such as freshwater, fracture proppant, or drilling rigs capable of drilling wells two miles or more in length.

1.3 RESERVOIRS

In conventional gas reservoirs (GAO, 2012; Speight, 2007, 2014), oil and gas are fairly mobile and easily move through the permeable formation because of buoyancy (they are lighter than the water in the same formation and therefore rise) until they are trapped against an impermeable rock (i.e., a seal) that prevents further movement. This leads to localized pools of oil and gas while the rest of the formation is filled with water. However, both biogenic and thermogenic shale gases remain where they were first generated and can be found in three forms: (i) free gas in the pore spaces and fractures; (ii) adsorbed gas, where the gas is electrically stuck to the organic matter and clay; and (iii) a small amount of dissolved gas that is dissolved in the organic matter.

In such reservoirs, typically an impermeable shale formation is either the basement rock or the cap rock of a sandstone formation thereby preventing any fluids within the sandstone from escaping or migrating to other formations.

A shale gas reservoir (*gas shale*) is an organic rich, and fine-grained shale that contains natural gas (Bustin, 2006; Bustin et al., 2008). However, the term *shale* is used very loosely and does not describe the lithology of the reservoir. Lithological variations in American shale gas reservoirs indicate that natural gas is retained in the reservoir not only in shale but also a wide spectrum of lithology and texture from mudstone (i.e., nonfissile shale) to siltstone and fine-grained sandstone, any of which may be of siliceous or carbonate composition. For example, in many basins, much of what is described as shale is often siltstone, or encompasses multiple rock types, such as siltstone or sandstone laminations interbedded with shale laminations or beds.

The presence of multiple rock types in organic-rich shale formations implies that there are multiple gas storage mechanisms, as gas may be adsorbed on organic matter and stored as free gas in micropores and macropores. Laminations serve a dual purpose because they both store free gas and transmit gas desorbed from organic matter in shale to the wellbore. The determination of the permeability and porosity of the laminations, and the linking of these laminations via a hydraulic fracture to the wellbore, are key requirements for efficient development. Additionally, solute or solution gas may be held in micropores and nanopores of bitumen (Bustin, 2006) and may be an additional source of gas, although traditionally this is thought to be a minor component. Free gas may be a more dominant source of production than desorbed gas or solute gas in a shale gas reservoir. Determining the percentage of free gas versus solute gas versus desorbed gas is important for resource and reserve evaluation and is a significant issue in gas production and reserve calculations, as desorbed gas diffuses at a lower pressure than free gas.

The lack of a strict definition for shale causes an additional degree of difficulty for resource evaluation. Such a broad spectrum of lithology appears to form a transition with other resources, such as *tight gas,* where the difference between the tight gas reservoir and gas shale may be the higher amount of sandstone (in the tight gas reservoir) and the tight gas reservoir may actually contain no organic matter.

Briefly, a *shale gas reservoir* (*shale gas play*) is similar on a world-wide basis insofar as organic rich, gas-prone shale is generally difficult to *discover*. In all cases, a thorough understanding of the fundamental geochemical and geological attributes of shale is essential for resource assessment, development, and environmental stewardship. Four properties that are important characteristics in each shale gas play are: (i) the maturity of the organic matter, (ii) the type of gas generated and stored in the reservoir—biogenic gas or thermogenic gas, (iii) the total organic carbon (TOC) content of the strata, and (iv) the permeability of the reservoir.

A *tight gas reservoir* (*tight gas sands*) is a low-permeability sandstone reservoir that produces primarily dry natural gas. A tight gas reservoir is one that cannot be produced at economic flow rates or recover economic volumes of gas unless the well is stimulated by a large hydraulic fracture treatment and/or produced using horizontal wellbores. This definition also applies to coalbed methane (CBM) and tight carbonate reservoirs—shale gas reservoirs are also included by some observers (but not in this text).

In addition, the variety of rock types observed in organic-rich shale formations implies the presence of a range of different types of shale gas reservoirs. Each reservoir may have distinct geochemical and geological characteristics that may require equally unique methods of drilling, completion, production, and resource and reserve evaluation (Cramer, 2008)—leading to further necessary considerations when the shale gas had to be processed (see Chapter 4). Additionally, it must not be forgotten that a shale formation is often a seal or cap rock for a conventional (sandstone) petroleum or natural gas reservoir and that not all shale is necessarily reservoir rock (Speight, 2007, 2014).

On a more physical note, typical shale formations can be anywhere from 20 ft to a mile or so thick and extend over very wide geographic areas. A gas shale reservoir is often referred to as a *resource play*, where natural gas resources are widely distributed over extensive areas (perhaps several fields) rather than concentrated in a specific location. The volume of natural gas contained within a resource play increases as the thickness and areal extent of the deposit grows. Individual gas shale formations may have a billion cubic feet (1×10^9 ft^3) or even a trillion cubic feet (1×10^{12} ft^3) of gas in place spread over hundreds to thousands of square miles. The difficulty lies in extracting even a small fraction of that gas.

Shale formations exhibit a wide range of mechanical properties and significant anisotropy reflecting their wide range of material composition and fabric anisotropy (Sone, 2012). The elastic properties of these shale rocks are successfully described by tracking the relative amount of soft components (clay and solid organic materials) in the rock and also acknowledging the anisotropic distribution of the soft components. Gas shale formations also possess relatively stronger degree of anisotropy compared to other organic-rich shale formations, possibly due to the fact that these rocks come from peak-maturity source rocks. The deformational properties are influenced by the amount of soft components in the rock and exhibit mechanical anisotropy.

The pore spaces in shale, through which the natural gas must move if the gas is to flow into any well, are as much as 1000 times smaller than pores in conventional sandstone reservoirs. The gaps that connect pores (the pore throats) are smaller still, only 20 times larger than a single methane molecule. Therefore, shale has very low permeability but natural or induced fractures, which act as conduits for the movement of natural gas, will increase the permeability of the shale.

There is also the possibility (only assiduous geological studies will tell) of hybrid gas shale formation, where the originally deposited mud was rich in sand or silt. These foreign minerals (sand silt) result in a natural higher permeability for the shale formation and result in greater susceptibility of the shale to hydraulic fracturing.

1.4 SHALE GAS

Shale gas is natural gas produced from shale formations that typically function as both the reservoir and the source rocks for the natural gas. In terms of chemical makeup, shale gas is typically a dry gas composed primarily of methane (60–95% v/v), but some formations do produce wet gas. The Antrim and New Albany plays have typically produced water and gas. Gas shale formations are organic-rich shale formations that were previously regarded only as source rocks and seals for gas accumulating in the strata near sandstone and carbonate reservoirs of traditional onshore gas development.

Shale is a sedimentary rock that is predominantly composed of consolidated clay-sized particles. Shale formations are deposited as muds in low-energy environments such as tidal flats and deepwater basins where the fine-grained clay particles fall out of suspension in the quiet waters. During the deposition of these very fine-grained sediments, there can also be accumulation of organic matter in the form of algae, plant, and animal-derived organic debris (Davis, 1992). The very fine sheet-like clay mineral grains and laminated layers of sediment result in a rock with permeability that is limited horizontally and extremely limited vertically. This low permeability means that gas trapped in shale cannot move easily within the rock except over geologic expanses of time, i.e., millions of years. These units are often organic rich and are thought to be the source beds for a large percentage of the hydrocarbons produced in these basins.

Thus, by definition, *shale gas* is the hydrocarbon gas present in organic rich, fine grained, sedimentary rocks (shale and associated lithofacies). The gas is generated and stored *in situ* in gas shale as both sorbed gas (on organic matter) and free gas (in fractures or pores). As such, shale containing gas is a self-sourced reservoir. Low-permeable shale requires extensive fractures (natural or induced) to produce commercial quantities of gas.

Shale is a very fine-grained sedimentary rock, which is easily breakable into thin, parallel layers. It is a very soft rock, but it does not disintegrate when it becomes wet. The shale formations can contain natural gas, usually when two thick, black shale deposits *sandwich* a thinner area of shale. Because of some of the properties of the shale deposits, the extraction of natural gas from shale formations is more difficult and perhaps more expensive than that of conventional natural gas. Shale basins are scattered across the United States.

There are several types of unconventional gas resources that are currently produced. They are as follows:

1. Deep natural gas—natural gas that exists in deposits very far underground, beyond "conventional" drilling depths, typically 15,000 ft or more.

2. Shale gas—natural gas that occurs in low-permeability shale formations.
3. Tight natural gas—natural gas that occurs in low-permeability formations.
4. Geopressurized zones—natural underground formations that are under unusually high pressure for their depth.
5. Coalbed methane—natural gas that occurs in conjunction with coal seams.
6. Methane hydrates—natural gas that occurs at low temperature and high pressure regions such as the seabed and is made up of a lattice of frozen water, which forms a *cage* around the methane.

CBM is produced from wells drilled into coal seams which act as source and reservoir to the produced gas (Speight, 2013). These wells often produce water in the initial production phase, as well as natural gas. Economic CBM reservoirs are normally shallow, as the coal matrix tends to have insufficient strength to maintain porosity at depth.

On the other hand, shale gas is obtained from ultra-low permeability shale formations that may also be the source rock for other gas reservoirs. The natural gas volumes can be stored in fracture porosity, within the micropores of the shale itself, or adsorbed onto the shale.

In the context of this book, the focus is on *shale gas* and, when necessary, reference will also be made to *tight gas*. In respect to the low permeability of these reservoirs, the gas must be developed via special techniques including stimulation by hydraulic fracturing (or *fraccing*, *fracking*) in order to be produced commercially.

To prevent the fractures from closing when the pressure is reduced, several tons of sand or other *proppant* is pumped down the well and into the pressurized portion of the hole. When the fracturing occurs millions of sand grains are forced into the fractures. If enough sand grains are trapped in the fracture, it will be propped partially open when the pressure is reduced. This provides an improved permeability for the flow of gas to the well.

It has been estimated that there is in the order of 750 trillion cubic feet (Tcf, 1×10^{12} ft^3) of technically recoverable shale gas resources in the United States; represents a large and very important share of the United States recoverable resource base and, in addition, by 2035 approximately 46% of the natural gas supply of the United States will come from shale gas (EIA, 2011).

Tight gas is a form of unconventional natural gas that is contained in a very low-permeability formation underground—usually hard rock or a sandstone or limestone formation that is unusually impermeable and non-porous (tight sand). In a conventional natural gas deposit, once drilled, the gas can usually be extracted quite readily and easily (Speight, 2007). Like shale gas reservoirs, tight gas reservoirs are generally defined as having low permeability (in many cases less than 0.1 mD) (Law and Spencer, 1993). Tight gas makes up a significant portion of the natural gas resource base—more than 21% v/v of the total recoverable natural gas in the United States is in tight formations and represents an extremely important portion of natural gas resources (GAO, 2012).

In tight gas sands (low-porosity sandstones and carbonate reservoirs.), gas is produced through wells and the gas arose from a source outside the reservoir and migrated into the reservoir over geological time. Some *tight gas reservoirs* have also been found to be sourced by underlying coal and shale formation source rocks, as appears to be the case in the *basin-centered gas accumulations.*

However, extracting gas from a tight formation requires more severe extraction methods—several such methods do exist that allow natural gas to be extracted, including hydraulic fracturing and acidizing. It has been projected that gas-containing formations with permeability as little as 1 nD may be economically productive with optimized spacing and completion of staged fractures to maximize yield with respect to cost (McKoy and Sams, 2007). Like all unconventional natural gas, the economic incentive must be there to encourage companies to extract this costly gas instead of more easily obtainable, conventional natural gas.

The focus of this book is on shale gas, i.e., gas that exists in shale formation and requires additional efforts (such as fracking) for recovery but when necessary reference will also be made to *tight gas.*

1.4.1 Origin

Gas from shale is generated in two different ways, although a mixture of gas types is possible: (i) thermogenic gas is generated from cracking of organic matter or the secondary cracking of oil, and (ii) biogenic gas, such as in the Antrim shale gas field in Michigan, is generated from microbes in areas of fresh water recharge (Martini et al., 1998, 2003, 2004; Shurr and Ridgley, 2002). Thermogenic gas is associated with mature organic matter that has been subjected to relatively high temperature and pressure in order to generate hydrocarbons. Broadly speaking, more mature organic matter should generate higher gas-in-place resources than less mature organic matter, all other factors being equal (Martini et al., 1998; Schettler and Parmely, 1990).

Gas from shale is generated in two different ways, although a mixture of gas types is possible: (i) thermogenic gas is generated from thermal decomposition of organic matter or the secondary thermal decomposition of any liquid products—oil, and (ii) biogenic gas, such as in the Antrim shale gas field in Michigan, is generated from microbes in areas of fresh water recharge (Martini et al., 1998, 2003, 2004). Thermogenic gas is associated with mature organic matter that has been subjected to relatively high temperature and pressure in order to generate hydrocarbons.

Generally, more mature organic matter should generate higher gas-in-place resources than less mature organic matter, all other factors being equal. Organic maturity is often expressed in terms of vitrinite reflectance (% Ro), where a value above approximately 1.0–1.1% indicates that the organic matter is sufficiently mature to generate gas and could be an effective source rock.

Well-fractured shale that typically contains an abundance of mature organic matter and is deep or under high pressure will yield a high initial flow rate. For example, horizontal wells in the Barnett with a high initial reservoir pressure can yield an initial flow rate of a few million cubic feet per day after induced fracturing. However, after the first year, gas flow may be dominated by the rate of diffusion from the matrix to the induced fractures (Bustin et al., 2008).

Biogenic gas can be associated with either mature or immature organic matter, and can add substantially to shale gas reserves. For example, the San Juan Basin CBM gas field is a mixture of both gases and has generated much of its gas from biogenic processes (Scott et al., 1994). Likewise, gas from the Antrim shale formation in the Michigan Basin is largely biogenic gas that has been generated in the last 10,000–20,000 years (Martini et al., 1998, 2003, 2004) and more than 2.4 Tcf has been produced as of 2006. A mixture of gases is suggested for the New Albany shale formation in the Illinois Basin (Wipf and Party, 2006) and is certainly possible in Alberta shale.

1.4.2 Shale Reservoirs

Shale gas is natural gas that is produced from a type of sedimentary rock derived from clastic sources often including mudstone or siltstone which is known as shale. Clastic sedimentary rocks are composed of fragments (clasts) of preexisting rocks that have been eroded, transported, deposited, and lithified (hardened) into new rocks. Shale deposits typically contain organic material which was laid down along with the rock fragments.

In areas where conventional resources are located, shale can be found in the underlying rock strata and can be the source of the hydrocarbons that have migrated upwards into the reservoir rock. Over time, as the rock matures, hydrocarbons are produced from the kerogen. These may then migrate, as either a liquid (petroleum) or a gas (natural gas), through existing fissures and fractures in the rock until they reach the earth's surface or until they become trapped by strata of impermeable rock. Porous areas beneath these traps collect the hydrocarbons in a conventional reservoir, frequently of sandstone.

Shale gas reservoirs generally recover less gas (from <5% to 20% v/v) relative to conventional gas reservoirs (approximately 50–90% v/v) (Faraj et al., 2004), although the naturally well-fractured Antrim shale may have a recovery factor as high as 50–60% v/v. More recently, there have been suggestions that the Haynesville shale in Louisiana may have a recovery factor as high as 30% (Durham, 2008). To increase the recovery factor, innovation in drilling and completion technology is paramount in low-permeability shale reservoirs. In the initial state of pool development, permeability "sweet spots" are often sought because they result in higher rates of daily production and increased recovery of gas compared to less permeable shale.

But these sweet spots are small, relative to the size of unconventional pools, so horizontal drilling and new completion techniques (such as "staged fracs" and "simultaneous fracs"; Cramer, 2008) were developed to improve economics both inside and outside of "sweet spots." The result is a significant increase in economically producible reserves and a substantial extension of the area of economically producible gas.

Shale gas resources differ from conventional natural gas resources insofar as the shale acts as both the source for the gas and the zone (the reservoir) in which the gas is trapped. The very low permeability of the rock causes the rock to trap the gas and prevent it from migrating toward the surface. The gas can be held in natural fractures or pore spaces, or can be adsorbed onto organic material. With the advancement of drilling and completion technology, this gas can be successfully exploited and extracted commercially as has been proven in various basins in North America.

Aside from permeability, the key properties of shale, when considering gas potential, are (i) total organic content and (ii) thermal maturity.

The *total organic content* is the total amount of organic material present in the rock, expressed as a percentage by weight. Generally, the higher the total organic content, the better the potential for hydrocarbon generation. The *thermal maturity* of the rock is a measure of the degree to which organic matter contained in the rock has been heated over time, and potentially converted into liquid and/or gaseous hydrocarbons.

Because of the special techniques required for extraction, shale gas can be more expensive than conventional gas to extract. On the other hand, the in-place gas resource can be very large given the significant lateral extent and thickness of many shale formations. However, only a small portion of a shale gas resource may be theoretically producible and is even less likely to be producible in a commercially viable manner. Therefore, a key determinant of the success of a shale play is whether, and how much, gas can be recovered to the surface and at what cost.

The gas storage properties of shale are quite different to those of conventional reservoirs. In addition to having gas present in the matrix system of pores similar to that found in conventional reservoir rocks, shale also has gas bound or adsorbed to the surface of organic materials in the shale. The relative contributions and combinations of free gas from matrix porosity and from desorption of adsorbed gas is a key determinant of the production profile of the well.

The amount and distribution of gas within the shale is determined by, among other things, the initial reservoir pressure, the petrophysical properties of the rock, and its adsorption characteristics. During production there are three main processes at play. Initial gas production is dominated by depletion of gas from the fracture network. This form of production declines rapidly due to limited storage capacity. After the initial decline rate stabilizes, the depletion of gas stored in the matrix becomes the primary process involved in production. The amount of gas held in the matrix is dependent on the particular properties of the shale reservoir, which can be hard to estimate. Secondary to this depletion process is desorption whereby adsorbed gas is released from the rock as pressure in the reservoir declines. The rate of gas production via the desorption process depends on there being a significant drop in reservoir pressure.

Pressure changes typically advance through the rock very slowly due to low permeability. Tight well spacing can therefore be required to lower the reservoir pressure enough to cause significant amounts of adsorbed gas to be desorbed.

The ultimate recovery of the gas in place surrounding a particular shale gas well can be in the order of 28—40% of the original volume in place whereas the recovery per conventional well may be as high as 60—80% v/v. The development of shale gas plays, therefore, differs significantly from the development of conventional resources. With a conventional reservoir, each well is capable of draining oil or gas over a relatively large area (dependent on reservoir properties). As such, only a few wells (normally vertical) are required to produce commercial volumes from the field. With shale gas projects, a large number of relatively closely spaced wells are required to produce large enough volumes to make the plays economic. As a result, many wells must be drilled in a shale play to drain the reservoir sufficiently—in the Barnett shale resource in the United States, the drilling density can exceed one well per 60 acres.

TOC is a fundamental attribute of gas shale and is a measure of organic richness. The TOC content, the thickness of organic shale, and organic maturity are key attributes that aid in determining the economic viability of a shale gas play. There is no unique combination or minimum amount of these factors that determines economic viability. The factors are highly variable between shale of different ages and can vary, in fact, within a single deposit or stratum of shale over short distances. At the low end of these factors, there is very little gas generated. At higher values, more gas is generated and stored in the shale (if it has not been expelled from the source rock), and the shale can be a target for exploration and production. However, the presence of sufficient quantities of gas does not guarantee economic success, since shale has very low permeability and the withdrawal of gas is a difficult proposition that depends largely upon efficient drilling and completion techniques.

Induced fracturing may occur many times during the productive life of a shale gas reservoir (Walser and Pursell, 2007). Shale, in particular, exhibits permeability lower than CBM or tight gas and, because of this, forms the source and seal of many conventional oil and gas pools. Hence, not all shale is capable of sustaining an economic rate of production. In this respect, permeability of the shale matrix is the most important parameter influencing sustainable shale gas production (Bennett et al., 1991a,b; Bustin et al., 2008; Davies and Vessell, 2002; Davies et al., 1991; Gingras et al., 2004; Pemberton and Gingras, 2005).

To sustain yearly production, gas must diffuse from the low-permeability matrix to induced or natural fractures. Generally, higher matrix permeability results in a higher rate of diffusion to fractures and a higher rate of flow to the wellbore (Bustin et al., 2008). Furthermore, more fractured shale (i.e., shorter fracture spacing), given sufficient matrix permeability, should result in higher production rates (Bustin et al., 2008), a greater recovery of hydrocarbons, and a larger drainage area (Cramer, 2008; Walser and Pursell, 2007). Additionally, microfractures within shale matrix may be important for economic production; however, these microfractures are not easily determined *in situ* in a reservoir (Tinker and Potter, 2007), and only further research and analysis will determine their role in shale gas production.

An additional factor to consider is shale thickness. The substantial thickness of shale is one of the primary reasons, along with a large surface area of fine-grained sediment and organic matter for adsorption of gas, that shale resource evaluations yield such high values. Therefore, a general rule is that thicker shale is a better target. Shale targets such as the Bakken oil play in the Williston Basin (itself a hybrid conventional–unconventional play), however, are less than 50 m thick in many areas and are yielding apparently economic rates of flow. The required thickness to economically develop a shale gas target may decrease as drilling and completion techniques improve, as porosity and permeability detection techniques progress in unconventional targets and, perhaps, as the price of gas increases. Such a situation would add a substantial amount of resources and reserves to the province.

1.5 SHALE GAS AND ENERGY SECURITY

Energy security is the continuous and uninterrupted availability of energy to a specific country or region. The security of energy supply performs a crucial role in decisions that are related to the formulation of energy policy strategies. The economies of many countries are dependent on the energy imports in the notion that their balance of payments is affected by the magnitude of the vulnerability that the countries have in crude oil and natural gas (Speight, 2011).

Energy security has been an on-again-off-again political issue in the United States since the first Arab oil embargo in 1973. Since that time, the speeches of various presidents and the Congress of the United States have continued to call for an end to the dependence on foreign oil and gas by the United States. The congressional rhetoric of energy security and energy independence continues but meaningful suggestions of how to address this issue remain few and far between.

The energy literature and numerous statements by officials of oil-and-gas-producing and oil-and-gas-consuming countries indicate that the concept of energy security is elusive. Definitions of energy security range from uninterrupted oil supplies to the physical security of energy facilities to support for biofuels and renewable energy resources. Historically, experts and politicians referred to *security of oil supplies* as *energy security*. Only recently policy makers started to include natural gas supplies in the portfolio of energy definitions.

The security aspects of natural gas are similar, but not identical, to those of oil. Compared with oil imports, natural gas imports play a smaller role in most importing countries—mainly because it is less costly to transport liquid crude oil and petroleum products than natural gas. Natural gas is transported by pipeline over long distances because of the pressurization costs of transmission; the need to finance the cost of these pipelines encourages long-term contracts that dampen price volatility.

The past decade has yielded substantial change in the natural gas industry. Specifically, there has been rapid development of technology allowing the recovery of natural gas from shale formations. Since 2000, rapid growth in the production of natural gas from shale formations in North America has dramatically altered the global natural gas market landscape. Indeed, the emergence of shale gas is perhaps the most intriguing development in global energy markets in recent memory.

Beginning with the Barnett shale in northeast Texas, the application of innovative new techniques involving the use of horizontal drilling with hydraulic fracturing has resulted in the rapid growth in production of natural gas from shale. Knowledge of the shale gas resource is not new as geologists have long known about the existence of shale formations, and accessing those resources was long held in the geology community to be an issue of technology and cost. In the past decade, innovations have yielded substantial cost reductions, making shale gas production a commercial reality. In fact, shale gas production in the United States has increased from virtually nothing in 2000 to over 10 billion cubic feet per day (bcfd, 1×10^9 ft^3 per day) in 2010, and it is expected to more than quadruple by 2040, reaching 50% or more of total US natural gas production by the decade starting in 2030.

Natural gas—if not disadvantaged by government policies that protect competing fuels, such as coal—stands to play a very important role in the US energy mix for decades to come. Rising shale gas production has already delivered large beneficial impacts to the United States. Shale gas resources are generally located in close proximity to end-use markets where natural gas is utilized to fuel industry, generate electricity, and heat homes. This offers both security of supply and economic benefits (Medlock et al., 2011).

The *Energy Independence and Security Act of 2007* (originally named the *Clean Energy Act of 2007*) is an Act of Congress concerning the energy policy of the United States. The stated purpose of the act is "to move the United States toward greater energy independence and energy security, to increase the production of clean renewable fuels, to protect consumers, to increase the efficiency of products, buildings, and vehicles, to promote research on and deploy greenhouse gas capture and storage options, and to improve the energy performance of the Federal Government, and for other purposes."

The bill originally sought to cut subsidies to the petroleum industry in order to promote petroleum independence and different forms of alternative energy. These tax changes were ultimately dropped after opposition in the Senate, and the final bill focused on automobile fuel economy, development of biofuels, and energy efficiency in public buildings and lighting.

It was, and still is, felt by many observers that there should have been greater recognition of the role that natural gas can play in energy security.

In fact, viewed from the perspective of the energy-importing countries as a whole, diversification in oil supplies has remained constant over the last decade while diversification in natural gas supplies has steadily increased. Given the increasing importance of natural gas in world energy use, this points to an increase in overall energy security (Cohen et al., 2011).

However, natural gas is an attractive fuel, and its attraction is growing because of its clean burning characteristics, compared to oil or coal, and because of its price advantage, on an energy equivalent basis, compared to oil. Accordingly, analysts predict significant future growth in natural gas consumption worldwide and growth in the trade of natural gas. Significant investments are being made to meet this future demand by bringing so-called *stranded gas* (including *shale gas*) to market.

Current trends suggest that natural gas will gradually become a global commodity with a single world market, just like oil, adjusted for transportation differences. The outcome of a global gas market is inevitable; once this occurs, the tendency will be toward a world price of natural gas, as with oil today, and the prices of oil and gas each will reach a global equivalence based on energy content (Deutch, 2010).

REFERENCES

Bennett, R.H., Bryant, W.R., Hulbert, M.H. (Eds.), 1991a. Microstructure of Fine-Grained Sediments: From Mud to Shale. Springer-Verlag, New York, NY.

Bennett, R.H., O'Brien, N.R., Hulbert, M.H., 1991b. Determinants of clay and shale microfabric signatures: processes and mechanisms. In: Bennett, R.H., Bryant, W.R., Hulbert, M.H. (Eds.), Microstructure of Fine-Grained Sediments: From Mud to Shale. Springer-Verlag, New York, NY, pp. 5–32.

Blatt, H., Tracy, R.J., 2000. Petrology: Igneous, Sedimentary, and Metamorphic. W.H. Freeman and Company, New York, NY.

Bustin, A.M.M., Bustin, R.M., Cui, X., 2008. Importance of fabric on the production of gas shales. SPE Paper No. 114167. Proceedings of the Unconventional Gas Conference, Keystone, Colorado, February 10–12.

Bustin, R.M. 2006. Geology report: where are the high-potential regions expected to be in Canada and the U.S.? Capturing opportunities in Canadian shale gas. Second Annual Shale Gas Conference, The Canadian Institute, Calgary, January 31–February 1.

Cohen, G., Joutz, F., Loungani, P., 2011. Measuring energy security: trends in the diversification of oil and natural gas supplies. IMF Working Paper WP/11/39. International Monetary Fund, Washington, DC.

Cramer, D.D., 2008. Stimulating unconventional reservoirs: lessons learned, successful practices, areas for improvement. SPE Paper No. 114172. Proceedings of the Unconventional Gas Conference, Keystone, CO. February 10–12.

Davies, D.K., Vessell, R.K., 2002. Gas production from non-fractured shale. In: Scott, E.D., Bouma, A.H. (Eds.), Depositional Processes and Characteristics of Siltstones, Mudstones and Shale. Society of Sedimentary Geology, GCAGS Siltstone Symposium 2002, vol. 52. GCAGS (Gulf Coast Association of Geological Societies) Transactions, pp. 177–202.

Davies, D.K., Bryant, W.R., Vessell, R.K., Burkett, P.J., 1991. Porosities, permeabilities, and microfabrics of Devonian shales. In: Bennett, R.H., Bryant, W.R., Hulbert, M.H. (Eds.), Microstructure of Fine-Grained Sediments: From Mud to Shale. Springer-Verlag, New York, NY, pp. 109–119.

Davis Jr., R., 1992. Depositional Systems: An Introduction to Sedimentology and Stratigraphy, second ed. Prentice Hall, New York, NY.

Deutch, J., 2010. Oil and Gas Energy Security Issues. Resource for the Future. National Energy Policy Institute, Washington, DC.

Durham, L., 2008. Louisiana play a company maker? Am. Assoc. Pet. Geol. Explor. July (pp. 18, 20, 36).

EIA, 2011. Review of Emerging Resources: US Shale Gas and Shale Oil Plays. Energy Information Administration, United States Department of Energy, Washington, DC (July).

Faraj, B., Williams, H., Addison, G., McKinstry, B., 2004. Gas Potential of Selected Shale Formations in the Western Canadian Sedimentary Basin. GasTIPS, Hart Energy Publishing, Houston, TX, vol. 10 (1), pp. 21–25.

GAO, 2012. Information on shale resources, development, and environmental and public health risks. Report No. GAO-12-732. Report to Congressional Requesters. United States Government Accountability Office, Washington, DC. September.

Gingras, M.K., Mendoza, C.A., Pemberton, S.G., 2004. Fossilized worm burrows influence the resource quality of porous media. Am. Assoc. Pet. Geol. Bull. 88 (7), 875–883.

Hunter, C.D., Young, D.M., 1953. Relationship of natural gas occurrence and production in Eastern Kentucky (Big Sandy Gas Field) to joints and fractures in Devonian bituminous shales. Am. Assoc. Pet. Geol. Bull. 37 (2), 282–299.

Law, B.E., Spencer, C.W., 1993. Gas in tight reservoirs—an emerging major source of energy. In: Howell, D.G. (Ed.), The Future of Energy Gasses. United States Geological Survey, pp. 233–252. (Professional Paper No. 1570.)

Martini, A.M., Walter, L.M., Budai, J.M., Ku, T.C.W., Kaiser, C.J., Schoell, M., 1998. Genetic and temporal relations between formation waters and biogenic methane: Upper Devonian Antrim shale, Michigan Basin, USA. Geochim. Cosmochim. Acta 62 (10), 1699–1720.

Martini, A.M., Walter, L.M., Ku, T.C.W., Budai, J.M., McIntosh, J.C., Schoell, M., 2003. Microbial production and modification of gases in sedimentary basins: a geochemical case study from a Devonian shale gas play, Michigan Basin. Am. Assoc. Pet. Geol. Bull. 87 (8), 1355–1375.

Martini, A.M., Nüsslein, K., Petsch, S.T., 2004. Enhancing microbial gas from unconventional reservoirs: geochemical and microbiological characterization of methane-rich fractured black shales. Final Report. Subcontract No. R-520, GRI-05/0023. Research Partnership to Secure Energy for America, Washington, DC.

McKoy, M.L., Sams, W.N., 2007. Tight Gas Reservoir Simulation: Modeling Discrete Irregular Strata-Bound Fracture Networks and Network Flow, Including Dynamic Recharge from the Matrix. Contract No. DE-AC21-95MC31346, Federal Energy Technology Center, United States Department of Energy, Morgantown, West Virginia.

Medlock III, K.B., Jaffe, A.M., Hartley, P.R., 2011. Shale Gas and US National Security. James A. Baker III Institute for Public Policy, Rice University, TX.

Mokhatab, S., Poe, W.A., Speight, J.G., 2006. Handbook of Natural Gas Transmission and Processing. Elsevier, Amsterdam, The Netherlands.

Pemberton, G.S., Gingras, M.K., 2005. Classification and characterization of biogenically enhanced permeability. Am. Assoc. Pet. Geol. Bull. 89, 1493–1517.

Schettler, P.D., Parmely, C.R., 1990. The measurement of gas desorption isotherms for Devonian shale. Gas Shales Technol. Rev. 7 (1), 4–9.

Scott, A.R., Kaiser, W.R., Ayers, W.B., 1994. Thermogenic and secondary biogenic gases, San Juan Basin, Colorado and New Mexico: implications for coalbed gas productivity. Am. Assoc. Pet. Geol. Bull. 78 (8), 1186–1209.

Shurr, G.W., Ridgley, J.R., 2002. Unconventional shallow gas biogenic systems. Am. Assoc. Pet. Geol. Bull. 86 (11), 1939–1969.

Sone, H., 2012. Mechanical Properties of Shale Gas Reservoir Rocks and Its Relation to the In-Situ Stress Variation Observed In Shale Gas Reservoirs. A dissertation submitted to the Department of Geophysics and the Committee on Graduate Studies of Stanford University in Partial Fulfillment of the Requirements for the Degree of Doctor of Philosophy. SRB Volume 128, Stanford University, Stanford, CA.

Speight, J.G., 2007. Natural Gas: A Basic Handbook. GPC Books, Gulf Publishing Company, Houston, TX.

Speight, J.G., 2011. An Introduction to Petroleum Technology, Economics, and Politics. Scrivener Publishing, Salem, MA.

Speight, J.G., 2013. The Chemistry and Technology of Coal, third ed. CRC Press, Taylor & Francis Group, Boca Raton, FL.

Speight, J.G., 2014. The Chemistry and Technology of Petroleum, fifth ed. CRC Press, Taylor & Francis Group, Boca Raton, FL.

Tinker, S.W., Potter, E.C., 2007. Unconventional gas research and technology needs. Proceedings. Society of Petroleum Engineers R&D Conference: Unlocking the Molecules. San Antonio, TX, April 26–27.

Walser, D.W., Pursell, D.A., 2007. Making mature shale gas plays commercial: process and natural parameters. Proceedings. SPE Paper No. 110127. Society of Petroleum Engineers, Eastern Regional Meeting, Lexington, October 17–19.

Wipf, R.A. and Party, J.M., 2006. Shale Plays—A US Overview. AAPG Energy Minerals Division Southwest Section Annual Meeting, May.

CHAPTER 2

Shale Gas Resources

2.1 INTRODUCTION

The organic-rich shale formations (shale gas formations) have become an attractive target in the United States and Canada (Table 2.1) because they represent a huge resource of natural gas and, in some cases, natural gas liquids. Multiple operators are currently leasing and evaluating gas shale properties throughout the United States. If the prospective gas shale formations can be economically developed, many thousands of wells will be drilled in these regions during the next decade.

As a result, natural gas production in the United States has grown significantly in recent years as improvements in horizontal drilling and hydraulic fracturing technologies have made it commercially viable to recover gas trapped in tight formations, such as shale and coal. The United States is now the number one natural gas producer in the world and, together with Canada, accounts for more than 25% of global natural gas production (BP Statistical Review of World Energy, June 2012, www.bp.com). Shale gas will play an ever-increasing role in this resource base and in the economic outlook of the United States (Bonakdarpour et al., 2011). Furthermore, production of shale gas is projected to increase to 49% of total gas production in the United States by 2035, up from 23% in 2010, highlighting the significance of shale gas in the future energy mix in the United States. Lower and less volatile prices for natural gas in the recent past reflect these new realities, with benefits for American consumers and the nation's competitive and strategic interests, including the revitalization of several domestic industries.

The first Barnett Shale gas production, by Mitchell Energy and Development Corp, took place in the Fort Worth Basin in 1981. Until Barnett Shale successes, it was believed that natural fractures had to be present in gas shale. A low-permeability gas shale play is presently viewed as a technological play. Advances in microseismic fracture mapping, 3D seismic, horizontal drilling, fracture stimulation, and

Table 2.1 Shale Gas Formations in the United States and Canada

Formation	Period	Location
Antrim Shale	*Late Devonian*	*Michigan Basin, Michigan*
Baxter Shale	*Late Cretaceous*	*Vermillion Basin, Colorado, Wyoming*
Barnett Shale	*Mississippian*	*Fort Worth and Permian basins, Texas*
Bend Shale	Pennsylvanian	Palo Duro Basin, Texas
Cane Creek Shale	Pennsylvanian	Paradox Basin, Utah
Caney Shale	*Mississippian*	*Arkoma Basin, Oklahoma*
Chattanooga Shale	*Late Devonian*	*Alabama, Arkansas, Kentucky, Tennessee*
Chimney Rock Shale	Pennsylvanian	Paradox Basin, Colorado, Utah
Cleveland Shale	Devonian	Eastern Kentucky
Clinton Shale	Early Silurian	Eastern Kentucky
Cody Shale	Cretaceous	Oklahoma, Texas
Colorado Shale	*Cretaceous*	*Central Alberta, Saskatchewan*
Conasauga Shale	*Middle Cambrian*	*Black Warrior Basin, Alabama*
Dunkirk Shale	Upper Devonian	Western New York
Duvernay Shale	*Late Devonian*	*West central Alberta*
Eagle Ford Shale	*Late Cretaceous*	*Maverick Basin, Texas*
Ellsworth Shale	Late Devonian	Michigan Basin, Michigan
Excello Shale	Pennsylvanian	Kansas, Oklahoma
Exshaw Shale	Devonian–Mississippian	Alberta, northeast British Columbia
Fayetteville Shale	*Mississippian*	*Arkoma Basin, Arkansas*
Fernie Shale	Jurassic	West central Alberta, northeast British Columbia
Floyd/Neal Shale	*Late Mississippian*	*Black Warrior Basin, Alabama, Mississippi*
Frederick Brook Shale	Mississippian	New Brunswick, Nova Scotia
Gammon Shale	Late Cretaceous	Williston Basin, Montana
Gordondale Shale	Early Jurassic	Northeast British Columbia
Gothic Shale	Pennsylvanian	Paradox Basin, Colorado, Utah
Green River Shale	Eocene	Colorado, Utah
Haynesville/ Bossier Shale	*Late Jurassic*	*Louisiana, east Texas*
Horn River Shale	*Middle Devonian*	*Northeast British Columbia*
Horton Bluff Shale	*Early Mississippian*	*Nova Scotia*
Hovenweep Shale	Pennsylvanian	Paradox Basin, Colorado, Utah
Huron Shale	Devonian	East Kentucky, Ohio, Virginia, West Virginia
Klua/Evie Shale	Middle Devonian	Northeast British Columbia

(Continued)

Table 2.1 (Continued)		
Formation	Period	Location
Lewis Shale	*Late Cretaceous*	*Colorado, New Mexico*
Mancos Shale	*Cretaceous*	*San Juan Basin, New Mexico, Uinta Basin, Utah*
Manning Canyon Shale	Mississippian	Central Utah
Marcellus Shale	*Devonian*	*New York, Ohio, Pennsylvania, West Virginia*
McClure Shale	Miocene	San Joaquin Basin, California
Monterey Shale	Miocene	Santa Maria Basin, California
Montney-Doig Shale	*Triassic*	*Alberta, northeast British Columbia*
Moorefield Shale	Mississippian	Arkoma Basin, Arkansas
Mowry Shale	Cretaceous	Bighorn and Powder River basins, Wyoming
Muskwa Shale	Late Devonian	Northeast British Columbia
New Albany Shale	*Devonian–Mississippian*	*Illinois Basin, Illinois, Indiana*
Niobrara Shale	*Late Cretaceous*	*Denver Basin, Colorado*
Nordegg/Gordondale Shale	Late Jurassic	Alberta, northeast British Columbia
Ohio Shale	*Devonian*	*East Kentucky, Ohio, West Virginia*
Pearsall Shale	*Cretaceous*	*Maverick Basin, Texas*
Percha Shale	Devonian–Mississippian	West Texas
Pierre Shale	*Cretaceous*	*Raton Basin, Colorado*
Poker Chip Shale	Jurassic	West central Alberta, northeast British Columbia
Queenston Shale	Ordovician	New York
Rhinestreet Shale	Devonian	Appalachian Basin
Second White Speckled Shale	Late Cretaceous	Southern Alberta
Sunbury Shale	Mississippian	Appalachian Basin
Utica Shale	*Ordovician*	*New York, Ohio, Pennsylvania, West Virginia, Quebec*
Wilrich/Buckinghorse/Garbutt/ Moosebar Shale	Early Cretaceous	West central Alberta, northeast British Columbia
Woodford Shale	*Devonian–Mississippian*	*Oklahoma, Texas*

Formations discussed in this text are shown in **italic and bold** *font.*

multiple fracturing stages, have all contributed to successful gas shale wells.

By the early part of the twenty-first century, the main gas resources to that point had been: Antrim Shale in the northern Michigan Basin;

Barnett Shale in the Fort Worth Basin, TX; Lewis Shale in the San Juan Basin; New Albany Shale in the Illinois Basin; and the Ohio Shale in the Appalachian Basin (GAO, 2012). More recent gas shale resources include the Woodford Shale in Oklahoma; Fayetteville Shale in Arkansas; Haynesville Shale in Louisiana; Marcellus Shale in the Appalachian Basin; Utica Shale in New York; and Eagle Ford Shale in Texas.

In fact, in many parts of the United States and Canada, a reexamination of old drilling records is opening up opportunities for the "rediscovery" of gas and oil resources that were passed over at an earlier time of lower resource prices and/or more limited recovery technology. This is especially true with natural gas, which in many instances was a *stranded* resource having little or no market value. Also until quite recently with improvements in recovery technology, natural gas in tight sand or shale reservoirs could not be produced at commercial rates. There are more than 50 shale gas resource for formations in the United States and Canada, some of which are older (known) shale formation and other which are more recent and new. Of these shale resources, the most prominent or most interesting of these (at the time of writing, 2013) are listed in Table 2.1.

Shale gas reserves in the United States are considerable and not concentrated in any particular area. The estimates place 482 trillion cubic feet (482×10^{12} ft^3) of technically recoverable shale gas resources in the lower 48 states with the largest portions in the Northeast (63% v/v), Gulf Coast (13% v/v), and Southwest regions (10% v/v), respectively. The largest shale gas resources (*plays*) are the Marcellus Shale (141 trillion cubic feet, 141×10^{12} ft^3), Haynesville Shale (74.7 trillion cubic feet, 74.7×10^{12} ft^3), and Barnett Shale (43.4 trillion cubic feet, 43.4×10^{12} ft^3). Activity in new shale resources has increased shale gas production in the United States from 388 billion cubic feet (388×10^9 ft^3) in 2000 to 4944 billion cubic feet (4944×10^9 ft^3) in 2010 (EIA, 2011a). This production potential has the ability to change the nature of the North American energy mix and the natural gas resource base could support supply for five or more decades at current or greatly expanded levels of use (NPC, 2011).

However, in addition to these data, there are indications from numbers recently released that the estimated shale gas resources for the continental United States doubled from 2010 to 2011 to approximately

862 trillion cubic feet (862×10^{12} ft^3) and from 2006 to 2010 annual shale gas production in the United States almost quintupled to 4.8 trillion cubic feet (from 1.0 to 4.8×10^{12} ft^3) (EIUT, 2012).

Finally, each of the gas shale basins is different and each has its unique set of exploration criteria and operational challenges. Because of these differences, the development of shale gas resources in each of these areas poses potential challenges to the surrounding communities and ecosystems. For example, the Antrim and New Albany Shale formations are shallower shale formations which produce significant volumes of formation water unlike most of the other gas shale formations.

The following shale formations are not listed in any particular order, other than alphabetical order for ease of location.

2.2 US RESOURCES

Conventional resources of natural gas (or for that matter, any fossil fuel) exist in discrete, well-defined subsurface accumulations (reservoirs), with permeability values greater than a specified lower limit. Such conventional gas resources can usually be developed using vertical wells, and generally yield the high recovery factors.

Briefly, permeability is a measure of the ability of a porous medium, such as that found in a hydrocarbon reservoir, to transmit fluids, such as gas, oil, or water, in response to a pressure differential across the medium. In petroleum engineering, permeability is usually measured in units of millidarcies (mD).

By contrast, unconventional resources are found in accumulations where permeability is low (<0.1 mD). Such accumulations include *tight* sandstone formations, coalbeds (coalbed methane, CBM) and shale formations (Figure 2.1). Unconventional resource accumulations tend to be distributed over a larger area than conventional accumulations and usually require advanced technology such as horizontal wells or artificial stimulation in order to be economically productive; recovery factors are much lower—typically of the order of 15–30% of the gas initially in place (GIIP).

The mature, organic-rich shale formations that serve as sources for gas and which have received considerable interest, have become an

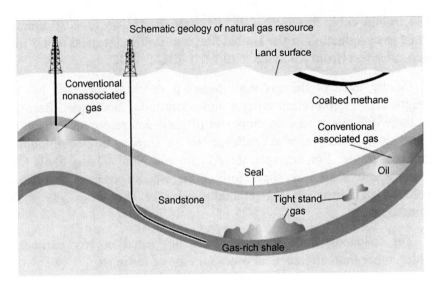

Figure 2.1 Illustration of the conventional and unconventional gas reservoirs. EIA (2011a).

attractive target because they represent a huge resource $(500-780 \times 10^{12} \text{ ft}^3)$, and are distributed throughout the 48 contiguous United States (Figure 2.2) (Hill and Nelson, 2000).

Due to the unique nature of shale, every basin, play, well, and pay zone may require a unique treatment. Briefly comparing the characteristics of some of the current hottest plays can help illustrate the impact of these differences throughout the development. It is necessary to study and understand the key reservoir parameters for gas shale deposits and these parameters include: (i) thermal maturity, (ii) reservoir thickness, (iii) total organic carbon (TOC) content, (iv) adsorbed gas fraction, (v) free gas within the pores and fractures, and (vi) permeability (see Chapter 4). The first two parameters are routinely measured. Thermal maturity is commonly measured in core analysis and reservoir thickness is routinely measured with logs. The calculation of the final four parameters requires a novel approach.

Almost all (96% v/v) of the shale natural gas in proved reserves in 2010 came from the six largest shale plays in the United States (EIA, 2012). The Barnett again ranked as the largest shale gas play in the United States, and significantly higher increases over 2009 proved reserves were registered by the Haynesville/Bossier (which more than doubled 2009 volumes) and the Marcellus (which nearly tripled).

Figure 2.2 Shale gas resources (shale gas plays) in the Contiguous United States. Adapted from EIA (2011a).

Among these six shale plays, the only decline from 2009 volumes was in the Antrim of northern Michigan—a mature, shallow biogenic shale gas play discovered in 1986 that is no longer being developed at the same pace as the other leading shale resources.

However, there are other gas shale resources that are of increasing importance to the United States energy balance and economics. These must not be ignored and the predominant shale gas resources in the United states are discussed below in *alphabetical order* and not in order of preference or importance.

2.2.1 Antrim Shale

The Antrim Shale (Michigan Basin) is part of an extensive, organic-rich shale depositional system that covered large areas of the ancestral North American continent in the Middle-to-Late Devonian. The intracratonic Michigan Basin was one of the several depocenters situated along the Eastern Interior Seaway. The basin has been filled with more than 17,000 ft of sediment, 900 ft of which comprises the Antrim Shale and associated Devonian–Mississippian rocks. The base of the Antrim, near the center of the modern structural basin, is approximately 2400 ft below sea level (Braithwaite, 2009; EIA, 2011a).

The Antrim Shale is a black, organic-rich bituminous shale which is divided into four members, from base to top: the Norwood, Paxton, Lachine, and upper members. The upper members are overlaid by the greenish-gray Ellsworth Shale.

The stratigraphy of the Antrim Shale is relatively straightforward and wells are typically completed in the Lachine and Norwood members of the lower Antrim, whose aggregate thickness approaches 160 ft. The TOC content of the Lachine and Norwood ranges from 0.5% to 24% w/w. These black shale formations are silica rich (20−41% microcrystalline quartz and wind-blown silt) and contain abundant dolomite and limestone concretions and carbonate, sulfide, and sulfate cements. The remaining lower Antrim unit, the Paxton, is a mixture of lime mudstone and gray shale lithology (Martini et al., 1998) containing 0.3−8% w/w TOC and 7−30% w/w silica. Correlation of the fossil alga *Foerstia* has established time equivalence among the upper part of the Antrim Shale, the Huron member of the Ohio Shale of the Appalachian Basin, and the Clegg Creek member of the New Albany Shale of the Illinois Basin (Roen, 1993).

Typical depths for the entire Antrim Shale unit range from 500 to 2300 ft and the areal extent is roughly approximately 30,000 square miles (Braithwaite, 2009; EIA, 2011a; Gutschick and Sandberg, 1991). The entire area is overlain by Devonian and Mississippian sediments and hundreds of feet of glacial till. The Antrim mineralogy shows the shale to be laminated with very fine grains. The composition consists mainly of illite and quartz with small quantities of organic material and pyrite.

The Antrim Shale has an organic matter content of up to 20% w/w and is mainly made up of algal material. The vitrinite reflectance is in the range of 0.4−0.6, indicating that the shale is thermally immature. The shale is also shallow and there is a high concentration of methane in the composition, which would lead one to assume the gas is of a microbial origin, but, $\delta^{13}C$ values indicate a more thermogenic origin (Martini et al., 1996).

For shallow wells in the Antrim Shale, the gas is of microbial origin. Deeper wells have a mix of thermogenic gas and microbial gas. For gas compositions with C1/(C2 + C3) <100 the gas origin is thermogenic, and this occurs for the gas present in the Niagaran Formation which underlays the Antrim Shale. Since the Antrim has so many natural fractures, it is reasonable to assume that there is migration of gas from the Niagaran Formation in to the Antrim Shale.

The Antrim Shale has two main ways of storing gas: absorption and free gas in the pore volume. The lower Norwood member has a higher adsorption capacity (approximately 115 ft^3 per ton) than the Lachine member (approximately 85 ft^3 per ton) (Kuuskraa et al., 1992). This is an important factor to consider when designing a fracture treatment because it would be more beneficial to have more of the proppant in the zone with the highest gas content. The free gas in the pore space can account for up to 10% of the total gas in place, but it is still not clear on how dependent the free gas is on the water in place. The very low permeability of the matrix could make it very difficult if not impossible to remove a significant portion of the free gas.

Two dominant sets of natural fractures have been identified in the northern producing trend, one oriented toward the northwest and the other to the northeast and both exhibiting subvertical to vertical inclinations. These fractures, generally uncemented or lined by thin coatings of calcite (Holst and Foote, 1981; Martini et al., 1998), have been mapped for several meters in the vertical direction and tens of meters horizontally in surface exposures. Attempts to establish production in the Antrim outside this trend have commonly encountered organic, gas-rich shale but with minimal natural fracturing and, hence, permeability (Hill and Nelson, 2000).

Thus, the Antrim Shale is highly fractured for a shale reservoir. Fracture spacing can be as close as 1–2 ft, compared to 10–20 ft for the Barnett Shale. These fractures can create permeability thicknesses in the range of 50–5000 md ft, which increases gas production. But, it also helps water flow, and thus most wells produce large amounts of water which must be disposed of (Kuuskraa et al., 1992).

2.2.2 Bakken Shale

The Bakken Shale of the Williston Basin of Montana and North Dakota has seen a similar growth rate to the Barnett. The Bakken is another technical play in which the development of this unconventional resource has benefitted from the technological advances in horizontal wells and hydraulic fracturing (Braithwaite, 2009; Cohen, 2008;

Cox et al., 2008). In April 2008, the United States Geological Survey (USGS) released an updated assessment of the undiscovered technically recoverable reserves for this shale play estimating there are 3.65 billion barrels (3.65×10^9 bbls) of oil, 1.85 trillion cubic feet (1.85×10^{12} ft^3) of associated natural gas, and 148 million barrels (148×10^6 bbls) of natural gas liquids in the play (EIA, 2011a; USGS, 2008).

The Bakken Shale formation differs from other shale plays in that it is an oil reservoir, a dolomite layered between two shale formations, with depths ranging from around 8000 to 10,000 ft from which oil, gas, and natural gas liquids are produced.

Each succeeding member of the Bakken Formation—lower shale, middle sandstone, and upper shale member—is geographically larger than the one below. Both the upper and lower shale formations, which are the petroleum source rocks, present fairly consistent lithology, while the middle sandstone member varies in thickness, lithology, and petrophysical properties.

The Bakken Shale formation is not as naturally fractured as the Barnett Shale formation and, therefore, requires more traditional fracture geometry with both longitudinal and transverse fractures. Diversion methods are used throughout hydraulic fracture treatments, which primarily use gelled water fracture fluids, although there is a growing trend toward the use of an intermediate strength proppant. Recently, the Bakken gas shale has seen an increase in activity, and the trend is toward longer laterals—up to 10,000 ft for single laterals in some cases. In addition, there is also a trend to drill below the lower Bakken Shale and fracture upward.

2.2.3 Barnett Shale
The wedge-shaped Fort Worth Basin covers approximately 15,000 square miles in North-Central Texas and is centered along the north–south direction, deepening to the north and outcropping at the Liano uplift in Liano County (Bowker, 2007a,b; EIA, 2011a; Jarvie et al., 2007). The Cambrian Riley and Hickory formations are overlaid by the Viola-Simpson and Ellenburger groups. The Viola-Simpson limestone group is found in Tarrant and Parker counties and acts as a barrier between the Barnett and the Ellenburger formations. The Ellenburger Formation is a very porous, karsted aquifer (Zuber et al., 2002) that if fractured will

produce copious amounts of highly saline water, effectively shutting down a well with water disposal cost.

Geochemical and reservoir parameters for the Barnett Shale in the Fort Worth Basin differ markedly from those of other gas-productive shales, particularly with respect to gas in place. For example, Barnett Shale gas is thermogenic in origin and hydrocarbon generation began in the Late Paleozoic, continued through the Mesozoic, and ceased with uplift and cooling during the Cretaceous (Jarvie et al., 2001, 2007). In addition, organic matter in the Barnett Shale formation has generated liquid hydrocarbons and Barnett-sourced oils in other formations, ranging from Ordovician to Pennsylvanian in age, in the western Fort Worth Basin (Jarvie et al., 2001, 2007)—cracking of this oil may have contributed to the gas-in-place resource.

The Mississippian age Barnett Shale overlies the Viola-Simpson group. The Barnett Shale varies in thickness from 150 to 800 ft and is the most productive gas shale in Texas. The permeability ranges from 7 to 50 nD and the porosity from 4% to 6% (Cipolla et al., 2010; Montgomery et al., 2005). In addition, well performance of the Barnett Shale changes significantly with changing produced fluid type, depth, and formation thickness (Hale and William, 2010) and depending on the type of completion method implemented and the large hydraulic fracture treatments (Ezisi et al., 2012).

The three most important production-related structures in the basin include both major and minor faulting, fracturing, and karst-related collapse features (Frantz et al., 2005). Fracturing is important to gas production because it provides a conduit for gas to flow from the pores to the wellbore, and it also increases exposure of the well to the formation. The Barnett Shale formation exhibits complex fracture geometry, which often creates difficulty in estimating fracture length and exposure to the formation due to the complex geometry. The fracturing is believed to be caused by the cracking of oil into gas. This cracking can cause a ten-fold increase in the hydrocarbon volume, increasing the pressure until the formation breaks. The precipitation of calcium carbonate in the fractures can cut down on the conductivity of the fractures. This precipitation is hard to detect on logs and can cause a well location that appears to be good on seismic to change into an unproductive well. This precipitation is also hard to treat with acidization, due to the long distances the acid is required to travel before making a noticeable impact on production.

Change in gas content with pressure occurs in the Barnett Shale with a typical reservoir pressure in the range of 3000–4000 psi (Frantz et al., 2005). In low-permeability formations, pseudo-radial flow can take over 100 years to be established. Thus, most gas flow in the reservoir is a linear flow from the near fracture area toward the nearest fracture face. Faulting and karst-related collapse features are important mainly in the Ellenburger Formation.

In addition to drilling longer laterals, current trends in the Barnett are toward bigger hydraulic fracturing projects and more stages. Infills are being drilled and testing of spacing is down to 10 acres, while refracturing of the first horizontal wells from 2003 to 2004 has commenced; both infills and refracturing are expected to improve the *estimated ultimate recovery* from 11% to 18% v/v. In addition, pad drilling (see Chapter 3), especially in urban areas, and recycling of water (see Chapter 5) are growing trends in the Barnett Shale, as elsewhere.

2.2.4 Baxter Shale

The Baxter Shale is stratigraphically equivalent to the Mancos, Cody, Steele, Hilliard, and Niobrara/Pierre formations (Braithwaite, 2009; EIA, 2011a; Mauro et al., 2010) and was deposited in hundreds of feet of water in the Western Interior Seaway from about 90 to 80 million years ago (Coniacean to lower Campanian) and consists of about 2500 ft of dominantly siliceous, illitic, and calcareous shale that contains regionally correlative coarsening-upward sequences of quartz- and carbonate-rich siltstones several tens of feet thick. The TOC content ranges from 0.5% to 2.5% in the shale and from 0.25% to 0.75% in the siltstones. Measured porosities in both the shale and the siltstones typically range from 3% to 6% with matrix permeability of 100–1500 nD.

Gas production has been established from the Baxter Shale in 22 vertical wells and 3 horizontal wells in the Vermillion Basin of northwestern Colorado and adjacent Wyoming. Production comes mainly from the silt-rich intervals as determined by production logs. The productive area in the Baxter Shale has vitrinite reflectance values approaching 2% and is in the dry gas generation window.

The resource area is defined by numerous wells with gas shows and over-pressuring in the Baxter Shale with gradients ranging from 0.6% to 0.8 psi/ft at depths greater than 10,000 ft.

A challenge within this reservoir is the ability to economically access this large unconventional gas accumulation. This is not a classic 100- to 300-foot-thick organic-rich shale gas reservoir. Instead it is a very large hydrocarbon resource stored in 2500 ft of shale with interbedded siltstone intervals. 3-D seismic data have proved useful in helping define potential fracture networks in the Baxter Shale that can be targeted with horizontal wells.

2.2.5 Caney Shale

The Caney Shale (Arkoma Basin, Oklahoma) is the stratigraphic equivalent of the Barnett Shale in the Fort Worth Basin (Andrews, 2007, 2012; Boardman and Puckette, 2006; Jacobi et al., 2009). The formation has become a gas producer since the large success of the Barnett Shale formation.

The Caney Shale, Chesterian age, was deposited in the Oklahoma part of the Arkoma Basin, one of a series of foreland basins that formed progressively westward along the Ouachita Fold Belt from the Black Warrior Basin in Mississippi to basins in southwest Texas. The Arkoma Basin in Oklahoma is in the southeast corner of the state north and northwest of the Ouachita Mountains.

The Caney Shale formation dips southward from a depth of 3000 ft in northern McIntosh county, Oklahoma to 12,000 ft north of the Choctaw thrust. The Caney Formation thickens toward the southeast from 90′ at its northwest edge to 220′ along the Choctaw fault in the south. It can be subdivided into six intervals based on characteristics of the density and resistivity logs.

Reported average TOC values for the Caney Formation range from 5% to 8% w/w, which show a linear correlation with density. Mud log gas values have a strong correlation with desorbed gas values and range from 120 to 150 ft^3/ton of shale. Estimates of gas in place for the Caney range from 30 to 40 billion cubic feet ($30-40 \times 10^9$ ft^3).

2.2.6 Chattanooga Shale

The Chattanooga Shale (Black Warrior Basin) has been considered as a rich oil shale formation (Rheams and Neathery, 1988). The Chattanooga sits within the thermogenic gas window in much of the Black Warrior Basin (Carroll et al., 1995) and may thus contain significant prospects for natural gas. The Chattanooga disconformably overlies Ordovician through Devonian strata, and the time value of the disconformity increases northward (Thomas, 1988). The Chattanooga is overlain sharply by the Lower Mississippian Maury Shale, which is commonly thinner than 2 ft, and the Maury is in turn overlain by the micritic Fort Payne Chert. The Chattanooga Shale in Alabama was apparently deposited in dysoxic to anoxic subtidal environments and can be considered as a cratonic extension of the Acadian foreland basin (Ettensohn, 1985).

The thickness of the Chattanooga varies significantly within the Black Warrior Basin. The shale is thinner than 10 ft and is locally absent in much of Lamar, Fayette, and Pickens counties, which is the principal area of conventional oil and gas production in the Black Warrior Basin. For this reason, the Chattanooga has not been considered to be the principal source rock for the conventional oil and gas reservoirs in this area. The shale is thicker than 30 ft in a belt that extends northwestward from Blount county into Franklin and Colbert counties. A prominent depocenter is developed along the southwestern basin margin in Tuscaloosa and Greene counties. Here, the shale is consistently thicker than 30 ft and is locally thicker than 90 ft.

The Chattanooga Shale is in some respects analogous to the Barnett Shale of the Fort Worth Basin in that it is an organic-rich black shale bounded by thick, mechanically stiff limestone units that may help confine induced hydrofractures within the shale (Gale et al., 2007; Hill and Jarvie, 2007). Because the Chattanooga is relatively thin, horizontal drilling combined with controlled hydrofracturing may maximize production rates.

2.2.7 Conasauga Shale

The Conasauga Shale gas formation continues to be developed primarily in northeast Alabama (EIA, 2011a). With the exception of one well in Etowah County and one well in Cullman County, all of the development has been in St. Clair County. Etowah and St. Clair counties are located northeast of Birmingham in the Valley and Ridge province of Alabama. Cullman County is north of Birmingham in the Cumberland Plateau province.

This shale formation represents the first commercial gas production from shale in Alabama, because it is geologically the oldest and most structurally complex shale formation from which gas production has been established. The Conasauga differs from other gas shale forma-tions in several respects. The productive lithology is thinly interbedded shale and micritic limestone that can contain more than 3% TOC.

The Conasauga Shale is of Middle Cambrian age and can be char-acterized as a shoaling-upward succession in which shale passes verti-cally into a broad array of inner ramp carbonate facies. The shale was deposited on the outer ramp, and the shale is thickest in basement gra-bens that formed during Late Pre-Cambrian to Cambrian Iapetan rift-ing (Thomas et al., 2000).

The shale facies of the Conasauga is part of the weak litho-tectonic unit that hosts the basal detachment of the Appalachian Thrust Belt in Alabama (Thomas, 2001; Thomas and Bayona, 2005). The shale has been thickened tectonically into anti-formal stacks that have been interpreted as giant shale duplexes, or mushwads (Thomas, 2001). In places, the shale is thicker than 8000 feet, and the shale is complexly folded and faulted at outcrop scale.

Surface mapping and seismic exploration reveal that at least three Conasauga anti-forms are preserved in the Alabama Appalachians. Exploration has focused primarily on the southeastern portion of the Gadsden anti-form, which is in St. Clair and Etowah counties. The Palmerdale and Bessemer anti-forms constitute the core of the Birmingham anticlinorium. The Palmerdale and Bessemer structures are overlain by a thin roof of brittle Cambrian–Ordovician carbonate rocks and Conasauga Shale facies are exposed locally. The Palmerdale structure is in the heart of the Birmingham metropolitan area and thus may be difficult to develop, whereas the southwestern part of the Bessemer structure is in rural areas and may be a more attractive exploration target. Additional thick shale bodies may be concealed below the shallow Rome thrust sheet in Cherokee and northeastern Etowah counties and perhaps in adjacent parts of Georgia (Mittenthal and Harry, 2004).

2.2.8 Eagle Ford Shale

The Eagle Ford Shale (discovered in 2008) is a sedimentary rock for-mation from the Late Cretaceous age underlying much of South Texas which covers 3000 square miles and consists of an organic-rich marine shale that has also been found to appear in outcrops (Braithwaite, 2009; EIA, 2011a).

This hydrocarbon-producing formation rich in oil and natural gas extends from the Texas–Mexico border in Webb and Maverick counties and extends 400 miles toward East Texas. The formation is 50 miles wide and an average of 250 ft thick at a depth between 4000 and 12,000 ft. The shale contains a high amount of carbonate, which makes it brittle and easier to apply hydraulic fracturing to produce the oil or gas.

The Eagle Ford Shale formation is estimated to have 20.81 trillion cubic feet (20.81×1012 ft^3) of natural gas and 3.351 billion barrels (3351×109 bbls) of oil.

2.2.9 Fayetteville Shale

With productive wells penetrating the Fayetteville Shale (Arkoma Basin) at depths between a few hundred and 7000 ft, this formation is somewhat shallower than the Barnett Shale formation (Braithwaite, 2009; EIA, 2011a). Mediocre production from early vertical wells stalled development in the vertically fractured Fayetteville, and only with recent introduction of horizontal drilling and hydraulic fracturing has drilling activity increased.

In the most active Central Fayetteville Shale, horizontal wells are drilled using oil-based mud in most cases and water-based mud in others. In addition, 3-D seismic is gaining importance as longer laterals of 3000-plus feet are drilled and more stages are required for hydraulic fracturing. With growing numbers of wells and a need for more infrastructure, pad drilling is another trend emerging in the Fayetteville.

2.2.10 Floyd Shale

The Upper Mississippian Floyd Shale is an equivalent of the prolific Barnett Shale of the Fort Worth Basin. The shale is an organic-rich interval in the lower part of the Floyd Shale that is informally called the Neal Shale, which is an organic-rich, starved-basin deposit that is considered to be the principal source rock for conventional hydrocarbons in the Black Warrior Basin.

The Floyd Shale is a black marine shale located stratigraphically below the Mississippian Carter sandstone and above the Mississippian Lewis sandstone (EIA, 2011a). Although the Carter and Lewis sandstones have historically been the most prolific gas-producing zones in the Black Warrior Basin region of Alabama, there has been no prior production history reported for the Floyd Shale. The Chattanooga Shale is located below the Floyd and is separated from it in most areas by the Tuscumbia Limestone and the Fort Payne Chert.

The Mississippian Floyd Shale is an equivalent of the prolific Barnett Shale of the Fort Worth Basin and the Fayetteville Shale of the Arkoma Basin and has thus been the subject of intense interest. The Floyd is a broadly defined formation that is dominated by shale and limestone and extends from the Appalachian Thrust Belt of Georgia to the Black Warrior Basin of Mississippi.

Usage of the term Floyd can be confusing. In Georgia, the type Floyd Shale includes strata equivalent to the Tuscumbia Limestone, and in Alabama and Mississippi, complex facies relationships place the Floyd above the Tuscumbia Limestone, Pride Mountain Formation, or Hartselle Sandstone and below the first sandstone in the Parkwood Formation. Importantly, not all Floyd facies are prospective as gas reservoirs. Drillers have long recognized a resistive, organic-rich shale interval in the lower part of the Floyd Shale that is called informally the Neal Shale (Cleaves and Broussard, 1980; Pashin, 1994). In addition to being the probable source rock for conventional oil and gas in the Black Warrior Basin, the Neal Shale has the greatest potential as a shale gas reservoir in the Mississippian section of Alabama and Mississippi. Accordingly, usage of the term Neal helps to specify the facies of the Floyd that contains prospective hydrocarbon source rocks and shale gas reservoirs.

2.2.11 Haynesville Shale
The Haynesville Shale (also known as the Haynesville/Bossier) is situated in the North Louisiana Salt Basin in northern Louisiana and eastern Texas with depths ranging from 10,500 to 13,500 ft (Braithwaite, 2009; EIA, 2011a; Parker et al., 2009). The Haynesville is an Upper Jurassic age shale bounded by sandstone (Cotton Valley Group) above and limestone (Smackover Formation) below.

The Haynesville Shale covers an area of approximately 9000 square miles with an average thickness of 200–300 ft. The thickness and areal extent of the Haynesville has allowed operators to evaluate a wider variety of spacing intervals ranging from 40 to 560 acres per well. Gas content estimates for the play are 100 to 330 scf/ton. The Haynesville Formation has the potential to become a significant shale gas resource for the United States with original gas-in-place estimates of 717 trillion cubic feet (717×10^{12} ft^3) and technically recoverable resources estimated at 251 trillion cubic feet (251×10^{12} ft^3).

Compared to the Barnett Shale, the Haynesville Shale is extremely laminated, and the reservoir changes over intervals as small as 4 in. to 1 ft. In addition, at depths of 10,500−13,500 ft, this play is deeper than typical shale gas formations creating hostile conditions. Average well depths are 11,800 feet with bottomhole temperatures averaging 155°C (300°F) and wellhead treating pressures that exceed 10,000 psi. As a result, wells in the Haynesville require almost twice the amount of hydraulic horsepower, higher treating pressures, and more advanced fluid chemistry than the Barnett and Woodford Shale formations.

The high-temperature range, from 125°C (260°F) to 195°C (380°F), creates additional problems in horizontal wells, requiring rugged, high-temperature/high-pressure logging evaluation equipment. The formation depth and high-fracture gradient demand long pump times at pressures above 12,000 psi. In deep wells, there is also concern about the ability to sustain production with adequate fracture conductivity. The use of large volumes of water for fracturing makes water conservation and disposal a primary issue.

The Bossier Shale, often linked with the Haynesville Shale, is a geological formation that produces hydrocarbon and delivers large amounts of natural gas when properly treated. While there is some confusion when distinguishing Haynesville Shale from the Bossier Shale, it is in fact a relatively simple comparison—the Bossier Shale lies directly above the Haynesville Shale but lies under the Cotton Valley sandstones. However, some geologists still consider the Haynesville Shale and the Bossier Shale to be the same.

The thickness of the Bossier Shale is approximately 1800 ft in the area of interest. The productive zone is located in the upper 500−600 ft of the shale. The Bossier Shale is located in eastern Texas and northern Louisiana.

The Upper Jurassic (Kimmeridgian to Lower Tithonian) Haynesville and Bossier Shale formations of East Texas and northwest Louisiana are currently one of the most important shale gas plays in North America, exhibiting overpressure and high temperature, steep decline rates, and resources estimated together in the hundreds of trillions of cubic feet. These shale gas resources have been studied extensively by companies and academic institutions within the last year, but to date the depositional setting, facies, diagenesis, pore evolution, petrophysics, best completion techniques, and geochemical characteristics of the Haynesville and Bossier shales are still poorly understood. Our work represents new insights into Haynesville and Bossier shale facies, deposition, geochemistry, petrophysics, reservoir quality, and stratigraphy in light of paleographic setting and regional tectonics.

Haynesville and Bossier shales deposition was influenced by basement structures, local carbonate platforms, and salt movement associated with the opening of the Gulf of Mexico Basin. The deep basin was surrounded by carbonate shelves of the Smackover/Haynesville Lime Louark sequence in the north and east and local platforms within the basin. The basin periodically exhibited restricted environment and reducing anoxic conditions, as indicated by variably increased molybdenum content, presence of framboidal pyrite, and TOC$-$S$-$Fe relationships. These organic-rich intervals are concentrated along and between platforms and islands that provided restrictive and anoxic conditions during Haynesville and part of Bossier times.

The mudrock facies range from calcareous-dominated facies near the carbonate platforms and islands to siliceous-dominated lithologies in areas where deltas prograded into the basin and diluted organic matter (e.g., northern Louisiana and northeast Texas). These facies are a direct response to a second-order transgression that lasted from the early Kimmeridgian to the Berriasian. Haynesville and Bossier shales each compose three upward-coarsening cycles that probably represent third-order sequences within the larger second-order transgressive systems and early highstand systems tracts, respectively. Each Haynesville third-order cycle is characterized by unlaminated mudstone grading into laminated and bioturbated mudstone. Most of the three Bossier third-order cycles are dominated by varying amounts of siliciclastic mudstones and siltstones. However, the third Bossier cycle exhibits higher carbonate and an increase in organic productivity in a southern restricted area (beyond the basinward limits of Cotton Valley progradation), creating another productive gas shale opportunity. This organic-rich Bossier cycle extends across the Sabine Island complex and the Mt. Enterprise Fault Zone in a narrow trough from Nacogdoches County, Texas, to Red River Parish, Louisiana. Similar to the organic-rich Haynesville cycles, each third-order cycle grades from unlaminated into laminated mudstone and is capped by bioturbated, carbonate-rich mudstone facies. Best reservoir properties are commonly found in facies with the highest TOC, lowest siliciclastics, highest level of maturity, and highest porosity. Most porosity in the Haynesville and Bossier is related to interparticle nano- and micropores and, to a minor degree, by porosity in organic matter.

Haynesville and Bossier gas shales are distinctive on wireline logs—high gamma ray, low density, low neutron porosity, high sonic traveltime, moderately high resistivity. A multimin log model seems to predict the TOC content from logs. Persistence of distinctive log signatures is similar for the organic-rich Bossier Shale and the Haynesville Shale across the study area, suggesting that favorable conditions for shale gas production extend beyond established producing areas.

2.2.12 Hermosa Shale

The black shale of the Hermosa Group (Utah) consists of nearly equal portions of clay-sized quartz, dolomite and other carbonate minerals, and various clay minerals. The clay is mainly illite with minor amounts of chlorite and mixed layer of chlorite–smectite (Hite et al., 1984).

The area of interest for the Hermosa Group black shale is the northeast half of the Paradox Basin, the portion referred to as the fold and fault belt. This is the area of thick halite deposits in the Paradox Formation, and consequently narrow salt walls and broad interdome depressions. To the southwest of this stratigraphically controlled structural zone the black shale intervals are fewer and thinner, and they lack the excellent seals provided by the halite. The area encompasses eastern Wayne and Emery counties, southern Grand County, and the northeast third of San Juan County (Schamel, 2005, 2006). The kerogen in the shale is predominantly gas-prone humic type III and mixed types II–III (Nuccio and Condon, 1996).

Numerous factors favor the possible development of shale gas in the black shale intervals of the Hermosa Group. First, the shales are very organic-rich, on the whole the most carbonaceous shale in Utah, and they are inherently gas-prone. Second, they have reached relatively high degrees of thermal maturity across much of the basin. Third and perhaps most significant, the shale is encased in halite and anhydrite which retard gas leakage, even by diffusion. Yet it is curious that the Paradox Basin is largely an oil province (Montgomery, 1992; Morgan, 1992) in which gas production is historically secondary and associated gas, which relates to the concentration of petroleum development in the shallower targets on the southwest basin margin and in the salt-cored anticlines.

2.2.13 Lewis Shale

The Lewis Shale (San Juan Basin) is a quartz-rich mudstone that was deposited in a shallow, offshore marine setting during an early Campanian transgression southwestward across shoreline deposits of the underlying progradational Cliffhouse Sandstone member of the Mancos Formation (EIA, 2011a; Nummendal and Molenaar, 1995). The gas resources of the Lewis Shale are currently being developed, principally through recompletions of existing wells targeting deeper, conventional sandstone gas reservoirs (Braithwaite, 2009; Dube et al., 2000).

The 1000–1500 ft thick Lewis Shale is lowermost shore-face and pro-delta deposits composed of thinly laminated (locally bioturbated) siltstones, mudstones, and shale. The average clay fraction is just 25%, but quartz is 56%. The rocks are very tight. Average matrix gas porosity is 1.7% and the average gas permeability is 0.0001 mD. The rocks also are organically lean, with an average TOC content of only 1.0%; the range is 0.5–1.6%. The reservoir temperature is 46°C (140°F). Yet the adsorptive capacity of the rock is 13–38 scf/ton, or about 22 billion cubic feet per quarter section (i.e., per 160 acres) (Jennings et al., 1997).

Four intervals and a conspicuous, basin-wide bentonite marker are recognizable in the shale. The greatest permeability is found in the lowermost two-thirds of the section, which may be the result of an increase in grain size and microfracturing associated with the regional north-south/eastwest fracture system (Hill and Nelson, 2000).

2.2.14 Mancos Shale

The Mancos Shale formation (Uinta Basin) is an emerging shale gas resource (EIA, 2011a). The thickness of the Mancos (averaging 4000 ft in the Uinta Basin) and the variable lithology present drillers with a wide range of potential stimulation targets. The area of interest for the Mancos Shale is the southern two-thirds of the greater Uinta Basin, including the northern parts of the Wasatch Plateau. In the northern one-third of the basin there have been two few well penetrations of the Mancos Shale, and it is too deep to warrant commercial exploitation of a "low density" resource such as shale gas. The area is within Duchesne, Uinta, Grand, Carbon, and the northern part of Emery counties (Braithwaite, 2009; Schamel, 2005, 2006).

The Mancos Shale is dominated by mudrock that accumulated in offshore and open-marine environments of the Cretaceous Interior seaway. It is 3450−4150 ft thick where exposed in the southern part of the Piceance and Uinta basins, and geophysical logs indicate the Mancos to be about 5400 ft thick in the central part of the Uinta Basin. The upper part of the formation is inter-tongued with the Mesaverde Group—these tongues typically have sharp basal contacts and gradational upper contacts. Named tongues include the Buck and the Anchor Mine Tongues. An important hydrocarbon-producing unit in the middle part of the Mancos was referred to as the Mancos B Formation, which consists of thinly interbedded and interlaminated, very fine-grained to fine-grained sandstone, siltstone, and clay that was interpreted to have accumulated as north-prograding fore slope sets within an open-marine environment. The Mancos B has been incorporated into a thicker stratigraphic unit identified as the Prairie Canyon member of the Mancos, which is approximately 1200 ft thick (Hettinger and Kirschbaum, 2003).

At least four members of the Mancos have shale gas potential: (i) the Prairie Canyon (Mancos B), (ii) the Lower Blue Gate Shale, (iii) the Juana Lopez, and (iv) the Tropic-Tununk Shale. Organic matter in the shale has a large fraction of Terrigenous material derived from the shorelines of the Sevier belt. The thickness of the organic-rich zones within individual system tracts exceeds 12 ft. Vitrinite reflectance values from a limited number of samples at the top of the Mancos range from 0.65% at the Uinta Basin margins to >1.5% in the central basin.

Across most of Utah, the Mancos Shale has not been sufficiently buried to have attained the levels of organic maturity required for substantial generation of natural gas, even in the humic kerogen-dominant (types II−III) shale that characterize this group (Schamel, 2005, 2006). However, vitrinite reflectance values beneath the central and southern

Uinta Basin are well within the gas generation window at the level of the Tununk Shale, and even the higher members of the Mancos Shale. In addition to the *in situ* gas within the shale, it is likely that some of the gas reservoir in the silty shale intervals has migrated from deeper source units, such as the Tununk Shale or coals in the Dakota.

The Mancos Shale warrants consideration as the significant gas reservoir and improved methods for fracture stimulation tailored to the specific rock characteristics of the Mancos lithology are required. The well completion technologies used in the sandstones cannot be applied to the shale rocks without some reservoir damage.

2.2.15 Marcellus Shale

The Marcellus Shale (Appalachian Basin), also referred to as the Marcellus Formation, is a Middle Devonian black, low density, carbonaceous (organic rich) shale that occurs in the subsurface beneath much of Ohio, West Virginia, Pennsylvania, and New York. Small areas of Maryland, Kentucky, Tennessee, and Virginia are also underlain by the Marcellus Shale (Braithwaite, 2009; Bruner and Smosna, 2011; EIA, 2011a).

The Marcellus Shale formations were 400 million years in the making, stretching from western Maryland to New York, Pennsylvania, and West Virginia and encompassing the Appalachian region of Ohio along the Ohio River. It has been estimated that the Marcellus Shale formation could contain as much as 489 trillion cubic feet of natural gas, a level that would establish the Marcellus as the largest natural gas resource in North America and the second largest in the world.

Throughout most of its extent, the Marcellus is nearly a mile or more below the surface. These great depths make the Marcellus Formation a very expensive target. Successful wells must yield large volumes of gas to pay for the drilling costs that can easily exceed a million dollars for a traditional vertical well and much more for a horizontal well with hydraulic fracturing. There are areas where the thick Marcellus Shale can be drilled at minimum depths and tend to correlate with the heavy leasing activity that has occurred in parts of northern Pennsylvania and western New York.

Natural gas occurs within the Marcellus Shale in three ways: (i) within the pore spaces of the shale, (ii) within vertical fractures (joints) that break through the shale, and (iii) adsorbed on mineral grains and organic material. Most of the recoverable gas is contained in the pore spaces. However, the gas has difficulty escaping through the pore spaces because they are very tiny and poorly connected.

The gas in the Marcellus Shale is a result of its organic content. Logic therefore suggests that the more organic material there is contained in the rock the greater its ability to yield gas. The areas with the greatest production potential might be where the net thickness of organic-rich shale within the Marcellus Formation is greatest. Northeastern Pennsylvania is where the thick organic-rich shale intervals are located.

The Marcellus Shale ranges in depth from 4000 to 8500 ft, with gas currently being produced from hydraulically fractured horizontal wellbores. Horizontal lateral lengths exceed 2000 feet, and, typically, completions involve multistage fracturing with more than three stages per well.

Before 2000, many successful natural gas wells had been completed in the Marcellus Shale. The yields of these wells were often unimpressive upon completion. However, many of these older wells in the Marcellus have a sustained production that decreases slowly over time and many of them continued to produce gas for decades. To exhibit the interest in this shale formation, the Pennsylvania Department of Environmental Protection reports that the number of drilled wells in the Marcellus Shale has been increasing rapidly. In 2007, 27 Marcellus Shale wells were drilled in the state; however, in 2011 the number of wells drilled had risen to more than 2000.

For new wells drilled with the new horizontal drilling and hydraulic fracturing technologies, the initial production can be much higher than what was seen in the old wells. Early production rates from some of the new wells have been over one million cubic feet of natural gas per day. The technology is so new that long-term production data is not available. As with most gas wells, production rates will decline over time, however, a second hydraulic fracturing treatment could stimulate further production.

2.2.16 Neal Shale
The Neal Shale is an organic-rich facies of the Upper Mississippian age Floyd Shale formation. The Neal Shale formation has long been recognized as the principal source rock that charged conventional

sandstone reservoirs in the Black Warrior Basin (Carroll et al., 1995; EIA, 2011a; Telle et al., 1987) and has been the subject of intensive shale gas exploration in recent years.

The Neal Shale is developed mainly in the southwestern part of the Black Warrior Basin and is in facies relationship with strata of the Pride Mountain Formation, Hartselle Sandstone, the Bangor Limestone, and the lower Parkwood Formation. The Pride Mountain−Bangor interval in the northeastern part of the basin constitutes a progradational parasequence set in which numerous stratigraphic markers can be traced southwestward into the Neal Shale. Individual parasequences tend to thin southwestward and define a clinoform stratal geometry in which near-shore facies of the Pride Mountain-Bangor interval pass into condensed, starved-basin facies of the Neal Shale.

The Neal Formation maintains the resistivity pattern of the Pride Mountain-Bangor interval, which facilitates regional correlation and assessment of reservoir quality at the parasequence level. The Neal Shale and equivalent strata were subdivided into three major intervals, and isopach maps were made to define the depositional framework and to illustrate the stratigraphic evolution of the Black Warrior Basin in Alabama. The first interval includes strata equivalent to the Pride Mountain Formation and the Hartselle Sandstone and thus shows the early configuration of the Neal Basin. The Pride Mountain-Hartselle interval contains barrier-strand plain deposits (Cleaves and Broussard, 1980; Thomas and Mack, 1982). Isopach contours define the area of the barrier-strand plain system in the northeastern part of the basin, and closely spaced contours where the interval is between 25 and 225 ft thick define a southwestward slope that turns sharply and faces southeastward in western Marion County. The Neal-starved basin is in the southwestern part of the map area, where this interval is thinner than 25 ft.

The second interval includes strata equivalent to the bulk of the Bangor Limestone. A generalized area of inner ramp carbonate sedimentation is defined in the northeastern part of the formation where the interval is thicker than 300 ft. Muddy, outer-ramp facies are concentrated where this interval thins from 300 to 100 ft, and the northeastern margin of the Neal-starved basin is marked by the 100 ft contour. Importantly, this interval contains the vast majority of the prospective Neal reservoir facies, and the isopach pattern indicates that the slope had prograded more than 25 miles southwestward during Bangor deposition.

The final interval includes strata equivalent to the lower Parkwood Formation. The lower Parkwood separates the Neal Shale and the main part of the Bangor Limestone from carbonate-dominated strata of the middle Parkwood Formation, which includes a tongue of the Bangor that is called the *Millerella* limestone. The Lower Parkwood is a succession of siliciclastic deltaic sediment that prograded onto the Bangor ramp in the northeastern part of the study area and into the Neal Basin in the southern part and contains the most prolific conventional reservoirs in the Black Warrior Basin (Cleaves, 1983; Mars and Thomas, 1999; Pashin and Kugler, 1992). The lower Parkwood is thinner than 25 ft above the inner Bangor ramp and includes a variegated shale interval containing abundant slickensides and calcareous nodules, which are suggestive of exposure and vertic soil formation. The area of deltaic sedimentation is where the lower Parkwood is thicker than 50 ft and includes constructive deltaic facies in the Neal Basin and destructive, shoal-water deltaic facies along the margin of the Bangor ramp. In the southern part of the study area, the 25 ft contour defines a remnant of the Neal Basin that persisted through lower Parkwood deposition. In this area, condensation of lower Parkwood sediment brings middle Parkwood carbonate rocks within 25 ft of the resistive Neal Shale.

2.2.17 New Albany Shale

The New Albany Shale (Illinois Basin) is organic-rich shale located over a large area in southern Indiana and Illinois and in Northern Kentucky (Braithwaite, 2009; EIA, 2011a; Zuber et al., 2002). The depth of the producing interval varies from 500 to 2000 ft depth, with thicknesses of approximately 100 ft. The shale is generally subdivided into four stratigraphic intervals: from top to bottom, these are (i) Clegg Creek, (ii) Camp Run/Morgan Trail, (iii) Selmier, and (iv) Blocher intervals.

The New Albany Shale can be considered to be a *mixed source rock* in which some parts of the basin produced thermogenic gas and other parts produced biogenic gas. This is indicated by the vitrinite reflectance in the basin, varying from 0.6 to 1.3 (Faraj et al., 2004). It is not known whether circulating ground waters recently generated this biogenic gas or whether it is original biogenic gas generated shortly after the time of deposition.

Most gas production from the New Albany comes from approximately 60 fields in northwestern Kentucky and adjacent southern Indiana. However, past and current production is substantially less than that from either the Antrim Shale or Ohio Shale. Exploration and development of the New Albany Shale was spurred by the spectacular development of the Antrim Shale resource in Michigan, but results have not been as favorable (Hill and Nelson, 2000).

Production of New Albany Shale gas, which is considered to be biogenic, is accompanied by large volumes of formation water (Walter et al., 2000). The presence of water would seem to indicate some level of formation permeability. The mechanisms that control gas occurrence and productivity are not as well understood as those for the Antrim and Ohio Shale formations (Hill and Nelson, 2000).

2.2.18 Niobrara Shale

The Niobrara Shale formation (Denver-Julesburg Basin, Colorado) is a shale rock formation located in northeast Colorado, northwest Kansas, southwest Nebraska, and southeast Wyoming. Oil and natural gas can be found deep below the earth's surface at depths of 3000–14,000 ft. Companies drill these wells vertically and even horizontally to get the oil and natural gas in the Niobrara Formation.

The Niobrara Shale is located in the Denver-Julesburg Basin, which is often referred to as the DJ Basin. This resource exciting oil shale play is being compared to the Bakken Shale resource, which is located in North Dakota.

2.2.19 Ohio Shale

The Devonian shale in the Appalachian Basin was the first produced in the 1820s. The resource extends from Central Tennessee to Southwestern New York and also contains the Marcellus Shale formation. The Middle and Upper Devonian shale formations underlie approximately 128,000 square miles and crop out around the rim of the basin. Subsurface formation thicknesses exceed 5000 ft and organic-rich black shale exceeds 500 ft (152 m) in net thickness (DeWitt et al., 1993).

The Ohio Shale (Appalachian Basin) differs in many respects from the Antrim Shale petroleum system. Locally, the stratigraphy is considerably more complex as a result of variations in depositional setting across the basin (Kepferle, 1993; Roen, 1993). The shale formations can be further subdivided into five cycles of alternating carbonaceous shale formations and coarser-grained clastic materials (Ettensohn, 1985). These five shale cycles developed in response to the dynamics of the Acadian orogeny and westward progradation of the Catskill delta.

The Ohio Shale, within the Devonian shale, consists of two major stratigraphic intervals: (i) the Chagrin Shale and (ii) the underlying Lower Huron Shale.

The Chagrin Shale consists of 700−900 ft of gray shale (Curtis, 2002; Jochen and Lancaster, 1993), which thins gradually from East to West. Within 100−150 ft, a transition zone consisting of interbedded black and gray shale lithology announces the underlying Lower Huron Formation. The Lower Huron Shale is 200−275 ft of dominantly black shale, with moderate amounts of gray shale and minor siltstone. Essentially all the organic matter contained in the lower Huron is thermally mature for hydrocarbon generation, based on vitrinite reflectance studies.

The vitrinite reflectance of the Ohio Shale varies from 1% to 1.3 %, which indicates that the rock is thermally mature for gas generation (Faraj et al., 2004). The gas in the Ohio Shale is consequently of thermogenic origin. The productive capacity of the shale is a combination of gas storage and deliverability (Kubik and Lowry, 1993). Gas storage is associated with both classic matrix porosity and gas adsorption onto clay and non-volatile organic material. Deliverability is related to matrix permeability although highly limited (10^{-9} to 10^{-7} mD) and a well-developed fracture system.

2.2.20 Pearsall Shale

The Pearsall Shale is a gas-bearing formation that garnered attention near the Texas−Mexico border in the Maverick Basin before development of the Eagle Ford Shale truly commenced. The Pearsall Shale formation is found below the Eagle Ford Formation at depths of 7000−12,000 ft with a thickness of 600−900 ft (Braithwaite, 2009).

The formation does have the potential to produce liquids east of the Maverick Basin. As of 2012, only a few wells had been drilled in the play outside of the Maverick Basin but early results indicate there is potential that has largely been overlooked.

2.2.21 Pierre Shale

The Pierre Shale, located in Colorado, produced two million cubic feet of gas in 2008. Drilling operators are still developing this rock formation, which lies at depths that vary between 2500 and 5000 ft, and will not know its full potential until more wells provide greater information about its limits (Braithwaite, 2009).

The Pierre Shale formation is a division of Upper Cretaceous rocks laid down from approximately 146 million to 65 million years ago and is named for exposures studied near old Fort Pierre, South Dakota. In addition to Colorado, the formation also occurs in South Dakota, Montana, Colorado, Minnesota, New Mexico, Wyoming, and Nebraska.

The formation consists of approximately 2000 ft of dark gray shale, some sandstone, and many layers of bentonite (altered volcanic-ash falls that look and feel much like soapy clays). In some regions, the Pierre Shale may be as little as 700 ft thick.

The lower Pierre Shale represents a time of significant changes in the Cretaceous Western Interior Seaway, resulting from complex interactions of tectonism and eustatic sea level changes. The recognition and redefinition of the units of the lower Pierre Shale has facilitated understanding of the dynamics of the basin. The Burning Brule member of the Sharon Springs Formation is restricted to the northern part of the basin and represents tectonically influenced sequences. These sequences are a response to rapid subsidence of the axial basin and the Williston Basin corresponding to tectonic activity along the Absoroka Thrust in Wyoming. Unconformities associated with the Burning Brule member record a migrating peripheral bulge in the Black Hills region corresponding to a single tectonic pulse on the Absoroka Thrust. Migration of deposition and unconformities supports an elastic model for the formation and migration of the peripheral bulge and its interaction with the Williston Basin (Bertog, 2010).

2.2.22 Utah Shale

There are five kerogen-rich shale units having reasonable potential for commercial development as shale gas reservoirs. These are (i) four members of the Mancos Shale in northeast Utah—the Prairie Canyon, the Juana Lopez, the Lower Blue Gate, and the Tununk, and (ii) the black shale facies within the Hermosa Group in southeast Utah (Schamel, 2005).

The Prairie Canyon and Juana Lopez members are both detached mudstone–siltstone–sandstone successions embedded within the Mancos Shale in northeast Utah. The Prairie Canyon member is up to 1200 ft thick, but the stratigraphically deeper Juana Lopez member is less than 100 ft thick. Both are similar in lithology and basin setting to the gas-productive Lewis Shale in the San Juan Basin. As in the Lewis Shale, the lean, dominantly humic kerogen is contained in the shale interlaminated with the siltstone–sandstone. The high quartz content is likely to result in a higher degree of natural fracturing than in the enclosing clay–mudstone rocks. Thus, these two may respond well to hydraulic fracturing. Also, the porosity of the sandstone interbeds averaging 5.4% can enhance gas storage. Both units extend beneath the southeast Uinta Basin reaching depths sufficient for gas generation and retention from the gas-prone kerogen. Although not known to be producing natural gas at present, both units are worthy of testing for add-on gas, especially in wells that are programmed to target Lower Cretaceous or Jurassic objectives.

The Lower Blue Gate and Tropic-Tununk shales generally lack the abundant siltstone–sandstone interbeds that would promote natural and induced fracturing, but they do have zones of observed organic richness in excess of 2.0% that might prove to be suitable places for shale gas where the rocks are sufficiently buried beneath the southern Uinta Basin and perhaps parts of the Wasatch Plateau.

The black shale facies in the Hermosa Group of the Paradox Basin is enigmatic. These shale formations contain mixed types II–III kerogen that should favor gas generation, yet oil with associated gas dominate current production. They are relatively thin, just a few tens of feet thick on an average, yet they are encased in excellent sealing rocks, salt, and anhydrite. In the salt walls (anticlines), the shale formations are complexly deformed making them difficult to develop even with directional drilling methods, but where they are likely less deformed in the interdome areas (synclines) they are very deep. Yet in these deep areas one can expect peak gas generation. The shale formations are overpressured, which suggests generation currently or in the recent past. Prospects are good that shale gas reservoirs can be developed in the Paradox Basin, but it may prove to be technically and economically challenging (Schamel, 2005).

2.2.23 Utica Shale

The Utica Shale is a rock unit located approximately 4000–14,000 ft below the Marcellus Shale and has the potential to become an enormous natural gas resource. The boundaries of the deeper Utica Shale formation extend under the Marcellus Shale region and beyond. The Utica Shale encompasses New York, Pennsylvania, West Virginia, Maryland, and even Virginia. The Utica Shale is thicker than the Marcellus and has already proven its ability to support commercial gas production.

The geologic boundaries of the Utica Shale formation extend beyond those of the Marcellus Shale. The Utica Formation, which was deposited 40–60 million years ($40-60 \times 10^6$ years) before the Marcellus Formation during the Paleozoic Era, is thousands of feet beneath the Marcellus Formation. The depth of Utica Shale in the core production area of the Marcellus Shale formation creates a more expensive environment in which to develop the Utica Shale formations. However, in Ohio the Utica Shale formation is as little as 3000 ft below the Marcellus Shale, whereas in sections of Pennsylvania the Utica Formation is as deep as 7000 ft below the Marcellus Formation creating a better economic environment to achieve production from the Utica Shale formation in Ohio. Furthermore, the investments in the infrastructure to extract natural gas from the Marcellus Shale formation also increase the economic efficiency of extracting natural gas from the Utica Shale.

Although the Marcellus Shale is the current unconventional shale drilling target in Pennsylvania, another rock unit with enormous potential lies a few thousand feet below the Marcellus.

The potential source rock portion of the Utica Shale is extensive and underlies portions of Kentucky, Maryland, New York, Ohio, Pennsylvania, Tennessee, West Virginia, and Virginia. It is also present beneath parts of Lake Ontario, Lake Erie, and part of Ontario, Canada. The geographic extent of the Utica Shale source rock along with the equivalent Antes Shale of central Pennsylvania and Point Pleasant Shale indicates an extremely large gas resource base. In keeping with this areal extent, the Utica Shale has been estimated to contain (at least) 38 trillion cubic feet (38×10^{12} ft^3) of technically recoverable natural gas (at the mean estimate) according to the first assessment of this continuous (unconventional) natural gas accumulation by the USGS.

In addition to natural gas, the Utica Shale is also yielding significant amounts of natural gas liquids and oil in the western portion of its extent and has been estimated to contain on the order of 940 million barrels (940×10^6 bbls) of unconventional oil resources and approximately 208 million barrels (208×10^6 bbls) of unconventional natural gas liquids. A wider estimate places gas resources of the Utica Shale to be from 2 trillion cubic feet to 69 trillion cubic feet ($2-69 \times 10^{12}$ ft^3), which put this shale on the same resource level as the Barnett Shale, the Marcellus Shale, and the Haynesville Shale formations.

2.2.24 Woodford Shale

The Woodford Shale, located in south central Oklahoma, ranges in depth from 6000 to 11,000 ft (Abousleiman et al., 2007; Braithwaite, 2009; EIA, 2011a; Jacobi et al., 2009). This formation is a Devonian age shale bounded by limestone (Osage Lime) above and undifferentiated strata below. Recent natural gas production in the Woodford Shale began in 2003 and 2004 with vertical well completions only. However, horizontal drilling has been adopted in the Woodford, as in other shale gas plays, due to its success in the Barnett Shale.

The Woodford Shale play encompasses an area of nearly 11,000 square miles. The Woodford play is in an early stage of development and is occurring at a spacing interval of 640 acres per well. The average thickness of the Woodford Shale varies from 120 to 220 ft across the play. The gas content in the Woodford Shale is higher on average than some of the other shale gas plays at 200−300 scf/ton. The original gas-in-place estimate for the Woodford Shale is similar to the Fayetteville Shale at 23 trillion cubic feet (23×10^{12} ft^3), while the technically recoverable resources are estimate at 11.4 trillion cubic feet (11.4×10^{12} ft^3).

Woodford Shale stratigraphy and organic content are well understood, but due to their complexity compared to the Barnett Shale, the formations are more difficult to drill and fracture. As in the Barnett, horizontal wells are drilled, although oil-based mud is used in the

Woodford Shale and the formation is harder to drill. In addition to containing chert and pyrite, the Woodford Formation is more faulted, making it easy to drill out of the interval; sometimes crossing several faults in a single wellbore is required.

Like the Barnett Shale, higher silica rocks are predominant in the best zones for fracturing in the Woodford Formation, although the Woodford has deeper and higher fracture gradients. Due to heavy faulting, 3-D seismic is extremely important, as the Woodford Formation trends toward longer laterals exceeding 3000 feet with bigger fracture projects and more stages. Pad drilling also will increase as the Woodford Shale formation continues expanding to the Ardmore Basin and to West Central Oklahoma in Canadian County.

2.3 WORLD RESOURCES

Significant amounts of shale gas occur outside of the United States in other countries. The initial estimate of technically recoverable shale gas resources in the 32 countries was 5760 trillion cubic feet $(5760 \times 10^{12} \text{ ft}^3)$ (EIA, 2011b). Adding the US estimate of the shale gas technically recoverable resources of 862 trillion cubic feet $(862 \times 10^{12} \text{ ft}^3)$ results in a total shale gas resource base estimate of 6622 trillion cubic feet $(6622 \times 10^{12} \text{ ft}^3)$ for the United States and the other 32 countries assessed. To put this shale gas resource estimate in context, the technically recoverable gas resources worldwide are approximately 16,000 trillion cubic feet $(16,000 \times 10^{12} \text{ ft}^3)$, largely excluding shale gas (EIA, 2011b). Thus, adding the identified shale gas resources to other gas resources increases total world technically recoverable gas resources by more than 40% to 22,600 trillion cubic feet $(22,600 \times 10^{12} \text{ ft}^3)$ (EIA, 2011b).

At a country level, there are two country groupings that emerge where shale gas development appears most attractive. The first group consists of countries that are currently highly dependent upon natural gas imports, have at least some gas production infrastructure, and their estimated shale gas resources are substantially relative to their current gas consumption. For these countries, shale gas development could significantly alter their future gas balance, which may motivate development. The second group consists of those countries where the shale gas resource estimate is large (>200 trillion cubic feet, $>200 \times 10^{12} \text{ ft}^3$)

and there already exists a significant natural gas production infrastructure for internal use or for export. Existing infrastructure would aid in the timely conversion of the resource into production, but could also lead to competition with other natural gas supply sources. For an individual country, the situation could be more complex.

The predominant shale gas resources are found in the countries listed alphabetically below.

2.3.1 Argentina (Neuquén Basin)

Argentina has 774 trillion cubic feet (774×10^{12} ft^3) of technically recoverable shale gas, making it the world's third-largest resource behind the United States and China. Located on Argentina's border with Chile, the Neuquén Basin is the largest source of hydrocarbons, holding 35% of the country's oil reserves and 47% of the gas reserves. Within the basin, the Vaca Muerta Shale formation may hold as much as 240 trillion cubic feet (240×10^{12} ft^3) of exploitable gas.

Argentina's biggest energy company, YPF, has found unconventional shale oil and natural gas in Mendoza province, confirming the extension of the massive Vaca Muerta area. Exploration at the Payun Oeste and Valle del Rio Grande blocks pointed to an estimated one billion barrels (1×10^9 bbls) of oil equivalent (boe) in unconventional oil and gas in Mendoza. Energy resources and reserves in the province, which border the Andes mountain range in western Argentina, currently stand at 685 million (685×10^6) barrels of oil equivalent.

2.3.2 Canada

Recent estimates (NEB, 2009) indicate that there is the potential for one quadrillion cubic feet (1×10^{15} ft^3) of gas in place in shale formation in Canada located in different areas but predominantly in the Western Canada Sedimentary Basin (WCSB) (Figure 2.3). However, high uncertainty, because gas shale formations are still in the initial stages of evaluation across Canada, precludes calculating more rigorous resource estimates for Canada at the current time (NEB, 2009).

2.3.2.1 Colorado Group

The Colorado Group consists of various shale-containing horizons deposited throughout southern Alberta and Saskatchewan globally during high sea levels of the middle Cretaceous, including the Medicine Hat and Milk River shale-containing sandstones, which have

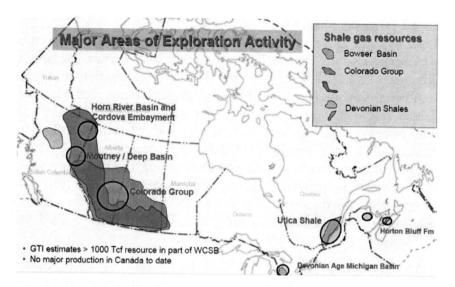

Figure 2.3 Occurrence of shale gas in Canada, especially the Western Canada Sedimentary Basin (WCSB). Adapted from NEB (2009).

been producing natural gas for over 100 years, and the Second White Speckled Shale, which has been producing natural gas for decades (Beaton et al., 2009).

In the Wildmere area of Alberta, the Colorado Shale is approximately 650 ft thick, from which natural gas has potential to be produced from five intervals. Unlike shale formations from the Horn River Basin and the Utica Group of Quebec, shale from the Colorado Group produces through thin sand beds and lamina, making it a hybrid gas shale like the Montney Shale. Furthermore, the gas produced in the Colorado has biogenic rather than thermogenic origins. This would suggest very low potential for natural gas liquids and an underpressured reservoir, which is more difficult to hydraulically fracture. Colorado Group shale formations are sensitive to water, which makes them sensitive to fluids used during hydraulic fracturing.

The total volume of gas in the Colorado Group is very difficult to estimate given the wide lateral extent of the shale and variability of the reservoir and the absence of independent and publicly available analyses. However, there could be at least 100 trillion cubic feet $(100 \times 10^{12} \text{ ft}^3)$ of gas in place.

2.3.2.2 Duvernay Shale

The Devonian Duvernay Shale is an oil and natural gas field located in Alberta, Canada (in the Kaybob area) which extends into British Columbia. The Duvernay Formation (Devonian–Frasnian) of Alberta, Canada is a Type II marine, proven source rock which has yielded much of the oil and gas to the adjacent classical Devonian, conventional fields in carbonate reefs and platform carbonates. Production in these conventional fields is in decline and exploration and development has now shifted to their source, the Duvernay Shale. The Devonian is considered the source rock for the Leduc reefs light oil resources, the discovery of which in 1947 was one of the defining moments in the past, present, and future Western Canadian oil and gas industry.

The Duvernay Shale, which can be found just north of the Montney Shale, is distributed over most of central Alberta and absent in areas of Leduc reef growth, except beneath the Duhamel reef, where it may be represented by a thin development of calcilutite (a dolomite or limestone formed of calcareous rock flour that is typically nonsiliceous). At its type section in the East Shale Basin, it is the shale formation of 174 ft thick—it thickens to 246 ft east and southeastward toward the Southern Alberta Shelf. Northeastward the formation reaches 394 ft at its truncation in the subsurface at the pre-Cretaceous unconformity. In the West Shale Basin, it averages 197 ft thick and thickens northward, attaining over 820 ft to the east of Lesser Slave Lake.

The formation consists of interbedded dark brown bituminous shale sediments, dark brown, black, and occasionally gray-green calcareous shale sediments and dense argillaceous limestone sediments. The shale formations are characteristically petroliferous and exhibit plane parallel millimeter lamination.

Based on petrophysics calibrated to core and cuttings samples, the Duvernay is characterized by porosity of 6.0–7.5%, permeability of 236–805 nD, and TOC content of 2.0–7.5% w/w. X-ray diffraction results from core and cuttings samples indicate it is likely very brittle with a low clay content (26% w/w), amorphous biogenic silica (47% w/w), and a calcite and dolomite matrix 20% w/w (Fowler et al., 2003; Switzer et al., 1994).

2.3.2.3 Horn River Basin

Devonian Horn River Basin shale formations were deposited in deep waters at the foot of the Slave Point carbonate platform in northeast British Columbia, which has been producing conventional natural gas for many decades. Horn River Basin shale formations are silica rich (approximately 55% v/v silica) and approximately 450 ft thick. The total organic content is 1−6%. The rocks are mature, having been heated far into the thermogenic gas window. The Horn River Shale formation located in British Columbia is the largest shale gas field in Canada and part of Canadian deposits that amount to as much as 250 trillion cubic feet (250×10^{12} ft^3) of natural gas (Ross and Bustin, 2008).

It should be noted that the Horn River Basin shale gas play also includes the Cordova Embayment and the whole formation extends into both the Yukon Territory and the Northwest Territories, although its northward extent beyond provincial/territorial borders is poorly defined.

2.3.2.4 Horton Bluff Group

Lacustrine muds of the Horton Bluff Group of the Canadian Maritime provinces were deposited in the Early Mississippian (approximately 360 million years ago) during regional subsidence (NEB, 2009). The silica content in the Frederick Brook Shale of the Horton Bluff Group in New Brunswick averages 38% v/v but the clay content is also high, averaging 42% v/v. There are indications that organic contents of the Frederick Brook member in Nova Scotia are significantly higher than in other Canadian gas shale formations, at 10% v/v, and the pay zone appears to be over 500 ft thick, sometimes exceeding 2500 ft in New Brunswick.

There are also indications that most of the gas is adsorbed onto clay and organic matter, and it will take very effective reservoir stimulation to achieve significant production from Nova Scotia Shale formations. It is unclear at this time what proportion of gas is adsorbed onto clay and organic matter in the New Brunswick Shale formations.

Analysis indicates that 67 trillion cubic feet (67×10^{12} ft^3) of free gas in place are present in the Frederick Brook Shale of the Sussex/Elgin subbasins of southern New Brunswick and 69 trillion cubic feet (69×10^{12} ft^3) of gas are present on the Windsor land block in Nova Scotia.

2.3.2.5 Montney Shale

The Montney Shale formation is a shale rock deposit located deep below British Columbia, Canada and is located in the Dawson Creek area just south of the Horn Rover Shale formation as well as the Duvernay Shale formation. Natural gas can be found in large quantities trapped in this tight shale formation (Williams and Kramer, 2011).

The formation is a hybrid between a tight gas and a shale gas resource and the sandy mudstone formation dates back to the Triassic period and is located beneath the Doig Formation at depths ranging from 5500 to 13,500 ft and is up to 1000 ft thick in places. As such, the Montney Shale is poised to become one of the most significant shale gas resources in Canada.

However, complicating reservoir characterization is that the upper and lower Montney zones in the same area have different mineralogy, which affects the formation evaluation data. The lower Montney is especially difficult as conventional open hole logs have historically caused people to believe that the lower Montney to be very tight. While the lower Montney has less porosity than the upper Montney Formation, core data over the lower zone has porosity higher than expected (Williams and Kramer, 2011).

The gas shale formation play is estimated to contain up to 50 trillion cubic feet (50×10^{12} ft^3) of natural gas trapped within poor permeability shale and siltstone. Horizontal wells are drilled at depths from 5500 to 13,500 ft and hydraulic fracturing enables the gas to flow more easily. Microseismic monitoring techniques can be used to assess fracture stimulations by locating events along each stage of the fracture and calculating the dimensions, geometry, and effective fracture volume. The Montney Shale is a unique resource play in that it is a hybrid between tight gas and traditional shale.

The Montney Formation is rich in silt and sand (characteristics similar to tight gas) but the source of the natural gas originated from its own organic matter like shale plays. Due to the presence of siltstone and sand, the Montney Formation has extremely low permeability and requires higher levels of fracture stimulation.

2.3.2.6 Utica Group

The Upper Ordovician Utica Shale is located between Montreal and Quebec City and was deposited in deep waters at the foot of the Trenton carbonate platform. Later, the shale was caught up in early Appalachian Mountain growth and became faulted and folded on its southeastern side. The Utica is approximately 500 ft thick and has a total organic content of 1–3% and has been known for decades as a petroleum source rock for associated conventional reservoirs.

However, unlike other Canadian gas shale formations, the Utica has higher concentrations of calcite, which occur at the expense of some silica (Theriault, 2008). While calcite is still brittle, hydraulic fractures do not transmit as well through it.

2.3.3 China (Sichuan and Tarim Basins)
In 2011, China was estimated to have 1275 trillion cubic feet $(1275 \times 10^{12} \text{ ft}^3)$ of technically recoverable shale gas. Since then a government geological survey confirmed a total of 882 trillion cubic feet $(882 \times 10^{12} \text{ ft}^3)$ of technically recoverable shale gas, excluding Tibet. The Sichuan Basin, located in south central China, accounts for 40% of the country's shale resources.

2.3.4 Poland
Eastern Europe may hold as much as 250 trillion cubic feet $(250 \times 10^{12} \text{ ft}^3)$ of shale gas, with the Silurian Shale gas resource holding as much as 187 trillion cubic feet $(187 \times 10^{12} \text{ ft}^3)$ of that total. These shale gas resources could reduce Europe's dependence on natural gas imports and will give Poland (Baltic–Podlasie–Lublin Basins) a strong claim to energy independence as its projected reserves equate to approximately 300 years of domestic consumption.

2.3.5 South Africa
In addition to being an area fertile in fossil remains, the Karoo Supergroup (South Africa) might also be one of the most plentiful sources of shale gas in the world. The area is constituted mainly of shale and sandstone and underlies more than two-thirds of the entire area of South Africa and contains an estimated 485 trillion cubic feet $(485 \times 10^{12} \text{ ft}^3)$ of technically recoverable gas. Shale gas could reduce the country's dependence on coal, to fuel 85% of the energy needs.

REFERENCES

Abousleiman, Y., Tran, M., Hoang, S., Bobko, C., Ortega, A., Ulm, F.J., 2007. Geomechanics field and laboratory characterization of Woodford Shale: the next gas play. Paper Bo. SPE 110120. Proceedings of the SPE Annual Technical Conference and Exhibition, November 11–14, Anaheim, CA.

Andrews, R.D., 2007. Stratigraphy, production, and reservoir characteristics of the Caney shale in Southern Oklahoma. Shale Shaker 58, 9–25.

Andrews, R.D., 2012. My favorite outcrop—Caney shale along the South Flank of the Arbuckle Mountains, Oklahoma. Shale Shaker 62, 273–276.

Beaton, A.P., Pawlowicz, J.G., Anderson, S.D.A., Rokosh, C.D., 2009. Rock Eval™ Total Organic Carbon, Adsorption Isotherms and Organic Petrography of the Colorado Group: Shale Gas Data Release. Energy Resources Conservation Board, Calgary, Alberta, Canada (Open File Report No. ERCB/AGS 2008–11).

Bertog, J., 2010. Stratigraphy of the lower Pierre shale (Campanian): implications for the tectonic and eustatic controls on facies distributions. J. Geol. Res. 2010, Article ID 910243, 15 pages, 2010. doi: 10.1155/2010/910243.   <http://www.hindawi.com/journals/jgr/2010/910243/cta/>   (accessed 09.05.2013.).

Boardman, D., Puckette, J., 2006. Stratigraphy and Paleontology of the Upper Mississippian Barnett Shale of Texas and Caney Shale of Southern Oklahoma. Oklahoma Geological Survey, Norman, OK (OGS Open-File Report No. 6-2006).

Bonakdarpour, M., Flanagan, R., Holling, C., Larson, J.W., 2011. The Economic and Employment Contributions of Shale Gas in the United States. Prepared for America's Natural Gas Alliance. IHS Global Insight (USA) Inc., Washington, DC, December.

Bowker, K.A., 2007a. Barnett shale gas production, Fort Worth Basin, issues and discussion. Am. Assoc. Pet. Geol. Bull. 91, 522–533.

Bowker, K.A., 2007b. Development of the Barnett Shale play, Fort Worth Basin. West Tex. Geol. Soc. Bull. 42 (6), 4–11.

Braithwaite, L.D., 2009. Shale-Deposited Natural Gas: A Review of Potential. Electricity Analysis Office, Electricity Supply Analysis Division, California Energy Commission, Sacramento, CA (Report No. CEC-200-2009-005-SD).

Bruner, K.R., Smosna, R., 2011. A comparative study of the Mississippian Barnett Shale, Fort Worth Basin, and Devonian Marcellus Shale, Appalachian Basin. United States Department of Energy, Morgantown Energy Technology Center, Morgantown, West Virginia (Report No. DOE/NETL-2011/1478).

Carroll, R.E., Pashin, J.C., Kugler, R.L., 1995. Burial History and Source-Rock Characteristics of Upper Devonian Through Pennsylvanian Strata, Black Warrior Basin, Alabama. Alabama Geological Survey, Tuscaloosa, AL (Circular No. 187).

Cipolla, C.L., Lolon, E.P., Erdle, J.C., Rubin, B., 2010. Reservoir modeling in Shale-Gas reservoirs (SPE Paper No. 125530). SPE Res. Eval. Eng. 13 (4), 638–653.

Cleaves, A.W., 1983. Carboniferous terrigenous clastic facies, hydrocarbon producing zones, and sandstone provenance, Northern Shelf of the Black Warrior Basin. Gulf Coast Assoc. Geol. Soc. Trans. 33, 41–53.

Cleaves, A.W., Broussard, M.C., 1980. Chester and Pottsville depositional systems, outcrop, and subsurface in the Black Warrior Basin of Mississippi and Alabama. Gulf Coast Assoc. Geol. Soc. Trans. 30, 49–60.

Cohen, D., 2008. An unconventional play in the Bakken. Energy Bull. April.

Cox, S.A., Cook, D.M., Dunek, K., Daniels, R., Jump. C., Barree, B., 2008. Unconventional Resource Play Evaluation: a look at the Bakken Shale Play of North Dakota. Paper No. SPE

114171. Proceedings of the SPE Unconventional Resources Conference, February 10–12, Keystone, CO.

Curtis, J.B., 2002. Fractured shale-gas systems. Am. Assoc. Pet. Geol. Bull. 86 (11), 1921–1938.

DeWitt Jr., W., Roen, J.B., Wallace, L.G., 1993. Stratigraphy of Devonian Black Shales and Associated Rocks in the Appalachian Basin. Bulletin No. 1909. Petroleum Geology of the Devonian and Mississippian Black Shale of Eastern North America. U.S. Geological Survey, pp. B1–B57.

Dube, H.G., Christiansen, G.E., Frantz Jr., J.H., Fairchild Jr., N.R., 2000. The Lewis Shale, San Juan Basin: What We Know Now. Society of Petroleum Engineers, Richardson, TX (SPE Paper No. 63091).

EIA, 2011a. Shale Gas and Shale Oil Plays. Energy Information Administration, United States Department of Energy, Washington, DC, July, <www.eia.gov>.

EIA, 2011b. World Shale Gas Resources: An Initial Assessment of 14 Regions Outside the United States. Energy Information Administration, United States Department of Energy, <www.eia.gov>.

EIA, 2012. U.S. Crude Oil, Natural Gas, and Natural Gas Liquids Proved Reserves, 2010. Energy Information Administration, United States Department of Energy, August, <www.eia.gov>.

EIUT, 2012. Fact-Based Regulation for Environmental Protection in Shale Gas Development Summary of Findings. The Energy Institute, University of Texas at Austin, Austin, TX, <http://energy.utexas.edu>.

Ettensohn, F.R., 1985. Controls on the Development of Catskill Delta Complex Basin-Facies. Special Paper No. 201. Geological Society of America, Boulder, CO, pp. 65–77.

Ezisi, L.B., Hale, B.W., William, M., Watson, M.C., Heinze, L., 2012. Assessment of Probabilistic Parameters for Barnett Shale Recoverable Volumes. SPE Paper No. 162915. Proceedings of the SPE Hydrocarbon, Economics, and Evaluation Symposium, September 24–25, Calgary, Canada.

Faraj, B., Williams, H., Addison, G., McKinstry, B., 2004. Gas Potential of Selected Shale Formations in the Western Canadian Sedimentary Basin. GasTIPS (Winter), Hart Energy Publishing, Houston, TX, pp. 21–25.

Fowler, M.G., Obermajer, M., Stasiuk, L.D., 2003. Rock-Eval and TOC Data for Devonian Potential Source Rocks, Western Canadian Sedimentary Basin. Geologic Survey of Canada, Calgary, Alberta, Canada (Open File No. 1579).

Frantz, J.H., Waters G.A., Jochen V.A., 2005. Evaluating Barnett Shale Production performance using an integrated approach. SPE Paper No. 96917. Proceedings of the SPE ATCE Meeting, October 9–12, Dallas, TX.

Gale, J.F.W., Reed, R.M., Holder, J., 2007. Natural fractures in the Barnett shale and their importance for hydraulic fracture treatments. Am. Assoc. Petrol. Geol. Bull. 91, 603–622.

GAO, 2012. Information on Shale Resources, Development, and Environmental and Public Health Risks. Report No. GAO-12-732. Report to Congressional Requesters. United States Government Accountability Office, Washington, DC. September.

Gutschick, R.C., Sandberg, C.A., 1991. Late Devonian history of the Michigan Basin. In: Catacosinos, P.A., Daniels, P.A. (Eds.), Early Sedimentary Evolution of the Michigan Basin. Geological Society of America, pp. 181–202. (Special Paper No. 256).

Hale, B.W., William, M., 2010. Barnett Shale: a resource play locally random and regionally complex. Paper No. SPE 138987. Proceedings of the SPE Eastern Regional Meeting, October 12–14, Morgantown, West Virginia.

Hettinger, R.D., Kirschbaum, M.A., 2003. Stratigraphy of the Upper Cretaceous Mancos Shale (Upper Part) and Mesaverde Group in the Southern Part of the Uinta and Piceance Basins, Utah and Colorado. In: Petroleum Systems and Geologic Assessment of Oil and Gas in the Uinta-Piceance

Province, Utah and Colorado. USGS Uinta-Piceance Assessment Team. U.S. Geological Survey Digital Data Series DDS−69−B. USGS Information Services, Denver Federal Center Denver, CO (Chapter 12).

Hill, D.G., Nelson, C.R., 2000. Gas Productive Fractured Shales: An Overview and Update. GasTIPS (Summer), Hart Energy Publishing, Houston, TX.

Hill, R.J., Jarvie, D.M. (Eds.), 2007. Barnett Shale. Am. Assoc. Pet. Geol. Bull. 91, 399−622.

Hite, R.J., Anders, D.E., Ging, T.G., 1984. Organic-Rich source rocks of Pennsylvanian age in the Paradox Basin of Utah and Colorado. In: Woodward, J., Meissner, F.F., Clayton, J.L. (Eds.), Hydrocarbon Source Rocks of the Greater Rocky Mountain Region. Guidebook, Rocky Mountain Association of Geologists Guidebook, Denver, CO, pp. 255−274.

Holst, T.B., Foote, G.R., 1981. Joint orientation in Devonian rocks in the Northern portion of the lower Peninsula of Michigan. Geol. Soc. Am. Bull. 92 (2), 85−93.

Jacobi, D., Breig, J., LeCompte, B., Kopal, M., Mendez, F., Bliven, S., et al., 2009. Effective geochemical and geomechanical characterization of Shale gas reservoirs from Wellbore environment: Caney and the Woodford Shale. Paper No. SPE 124231. Proceedings of the SPE Annual Technical Meeting, October 4−7, New Orleans, Louisiana.

Jarvie, D.M., Claxton, B.L., Henk, F., Breyer, J.T., 2001. Oil and shale gas from the Barnett Shale, Ft. Worth Basin, Texas. Proceedings of the AAPG Annual Meeting, pp. A100, Denver, Colorado.

Jarvie, D.M., Hill, R.J., Ruble, T.E., Pollastro, R.M., 2007. Unconventional shale-gas systems: the Mississippian Barnett shale of North Central Texas, as one model for thermogenic shale-gas assessment. Am. Assoc. Pet. Geol. Bull. 9, 475−499.

Jennings, G.L., Greaves, K.H., Bereskin, S.R., 1997. Natural Gas Resource Potential of the Lewis Shale, San Juan Basin, New Mexico and Colorado. Society of Petroleum Engineers, Richardson, TX (SPE Paper No. 9766).

Jochen, J.E., Lancaster, D.E., 1993. Reservoir characterization of an Eastern Kentucky Devonian Shale well using a naturally fractured, layered description. SPE Paper No. 26192. Proceedings of the SPE Gas Technology Symposium, June 28−30, Calgary, Alberta, Canada.

Kepferle, R.C., 1993. A depositional model and basin analysis for the gas-bearing black shale (Devonian and Mississippian) in the Appalachian Basin. In: Roen, J.B., Kepferle, R.C. (Eds.), Petroleum Geology of the Devonian and Mississippian Black Shale of Eastern North America. United States Geological Survey, Reston, VA, pp. F1−F23. (Bulletin No. 1909.)

Kubik, W., Lowry, P., 1993. Fracture identification and characterization using cores, FMS, CAST, and Borehole Camera: Devonian Shale, Pike County, Kentucky. SPE Paper No. 25897. Proceedings of the SPE Rocky Mountains Regional-Low Permeability Reservoirs Symposium, April 12−14, Denver, CO.

Kuuskraa, V.A., Wicks, D.E., Thurber, J.L., 1992. Geologic and reservoir mechanisms controlling gas recovery from the Antrim Shale. SPE Paper No. 24883. Proceedings of the SPE ATCE Meeting, October 4−7, Washington, DC.

Mars, J.C., Thomas, W.A., 1999. Sequential filling of a late Paleozoic Foreland Basin. J. Sediment. Res. 69, 1191−1208.

Martini, A.M., Budal, J.M., Walter, L.M., Schoell, N.M., 1996. Microbial generation of economic accumulations of methane within a shallow organic-rich shale. Nature 383 (6596), 155−158, September 12.

Martini, A.M., Walter, L.M., Budai, J.M., Ku, T.C.W., Kaiser, C.J., Schoell, M., 1998. Genetic and temporal relations between formation waters and biogenic methane: Upper Devonian Antrim Shale, Michigan Basin, USA. Geochim. Cosmochim. Acta 62 (10), 1699−1720.

Mauro, L., Alanis, K., Longman, M., Rigatti, V., 2010. Discussion of the Upper Cretaceous Baxter Shale gas reservoir, Vermillion Basin, Northwest Colorado and Adjacent Wyoming.

AAPG Search and Discovery Article #90122©2011. Proceedings of the AAPG Hedberg Conference, December 5–10, Austin, TX.

Mittenthal, M.D., Harry, D.L., 2004. Seismic Interpretation and Structural Validation of the Southern Appalachian Fold and Thrust Belt, Northwest Georgia, vol. 42. Georgia Geological Guidebook, University of West Georgia, Carrollton, GA, pp. 1–12.

Montgomery, S., 1992. Paradox Basin: cane creek play. Petrol. Front. 9, 66.

Montgomery, S.L., Jarvie, D.M., Bowker, K.A., Pollastro, R.M., 2005. Mississippian Barnett Shale, Fort Worth Basin, North-Central Texas: gas-shale play with multi-trillion cubic foot potential. Am. Assoc. Pet. Geol. Bull. 89 (2), 155–175.

Morgan, C.D., 1992. Horizontal drilling potential of the Cane Creek Shale, Paradox Formation, Utah. In: Schmoker, J.W., Coalson, E.B., Brown, C.A. (Eds.), Geological Studies Relevant to Horizontal Drilling: Examples from Western North America. Rocky Mountain Association of Geologists, pp. 257–265, Golden, Colorado.

NEB, 2009. A Primer for Understanding Canadian Shale Gas. National Energy Board, Calgary, Alberta, Canada, November.

NPC, 2011. Prudent Development: Realizing the Potential of North America's Abundant Natural Gas and Oil Resources. National Petroleum Council, Washington, DC, <www.npc.org>.

Nuccio, V.F., Condon, S.M., 1996. Burial and Thermal History of the Paradox Basin, Utah and Colorado, and Petroleum Potential of the Middle Pennsylvanian Paradox Formation. United States Geological Survey, Reston, Virginia (Bulletin No. 2000-O).

Nummendal, D., Molenaar, C.M., 1995. Sequence stratigraphy of ramp-setting strand plain successions: the Gallup Sandstone, New Mexico. In: Van Wagoner, J.C., Bertram G.T. (Eds.), Sequence Stratigraphy of Foreland Basin Deposits. AAPG Memoir No. 64, pp. 277–310.

Parker, M., Buller, D., Petre. E., Dreher, D., 2009. Haynesville Shale Petrophysical Evaluation. Paper No. SPE 122937. Proceedings of the SPE Rocky Mountain Petroleum Technology Conference, April 14–16, Denver, CO.

Pashin, J.C., 1994. Cycles and stacking patterns in carboniferous rocks of the Black Warrior Foreland Basin. Gulf Coast Assoc. Geol. Soc. Trans. 44, 555–563.

Pashin, J.C., Kugler, R.L., 1992. Delta-destructive spit complex in Black Warrior Basin: facies heterogeneity in Carter Sandstone (Chesterian), North Blowhorn Creek oil unit, Lamar county, Alabama. Gulf Coast Assoc. Geol. Soc. Trans. 42, 305–325.

Rheams, K.F., Neathery, T.L., 1988. Characterization and Geochemistry of Devonian Oil Shale, North Alabama, Northwest Georgia, and South-Central Tennessee (A Resource Evaluation). Alabama Geological Survey, Tuscaloosa, AL (Bulletin No. 128).

Roen, J.B., 1993. Introductory review—Devonian and Mississippian Black Shale, Eastern North America. In: Roen, J.B., Kepferle, R.C. (Eds.), Petroleum Geology of the Devonian and Mississippian Black Shale of Eastern North America. United States Geological Survey, Reston, Virginia, pp. A1–A8. (Bulletin No. 1909.)

Ross, D.K., Bustin, R.M., 2008. Characterizing the shale gas resource potential of Devonian–Mississippian Strata in the Western Canada sedimentary basin: application of an integrated formation evaluation. Am. Assoc. Pet. Geol. Bull. 92, 87–125.

Schamel, S., 2005. Shale Gas Reservoirs of Utah: Survey of an Unexploited Potential Energy Resource. Utah Geological Survey, Utah Department of Natural Resources, Salt Lake City, Utah (Open-File Report No. 461).

Schamel, S., 2006. Shale Gas Resources of Utah: Assessment of Previously Undeveloped Gas Discoveries. Utah Geological Survey, Utah Department of Natural Resources, Salt Lake City, Utah (Open-File Report No. 499).

Switzer, S.B., Holland, W.G., Christie, G.C., Graff, G.C., Hedinger, A.S., McAuley, R.J., et al., 1994. Devonian Woodbend-Winterburn Strata of the Western Canadian Sedimentary Basin. Geological Atlas of the Western Canadian Sedimentary Basin. CSPG/ARC, Calgary, Alberta, Canada, pp. 165–202 (Chapter 12).

Telle, W.R., Thompson, D.A., Lottman, L.K., Malone, P.G., 1987. Preliminary Burial–Thermal history investigations of the Black Warrior Basin: implications for Coalbed methane and conventional hydrocarbon development: Tuscaloosa, Alabama, University of Alabama. Proceedings 1987 Coalbed Methane Symposium, pp. 37–50.

Theriault, R., 2008. Caracterisation Geochimique Et Mineralogique et Evaluation du Potentiel Gazeifere des Shales De l'Utica et du Lorraine, Basses-Terres du Saint-Laurent. Quebec Exploration November 2008, Quebec City, Quebec.

Thomas, W.A., 1988. The Black Warrior Basin. In: Sloss, L.L. (Ed.), Sedimentary Cover—North American Craton, vol. D-2. The Geology of North America. Geological Society of America, Boulder, CO, pp. 471–492.

Thomas, W.A., 2001. Mushwad: Ductile duplex in the Appalachian thrust belt in Alabama. Am. Assoc. Petrol. Geol. Bull. 85, 1847–1869.

Thomas, W.A., Bayona, G., 2005. The Appalachian Thrust Belt in Alabama and Georgia: Thrust-Belt Structure, Basement Structure, and Palinspastic Reconstruction. Alabama Geological Society, Tuscaloosa, AL (Geological Survey Monograph No. 16).

Thomas, W.A., Mack, G.H., 1982. Paleogeographic relationship of a Mississippian Barrier-Island and Shelf-Bar system (Hartselle Sandstone) in Alabama to the Appalachian-Ouachita Orogenic Belt. Geol. Soc. Am. Bull. 93, 6–19.

Thomas, W.A., Astini, R.A., Osborne, W.E., Bayona, G., 2000. Tectonic framework of deposition of the Conasauga Formation. In: Osborne, W.E., Thomas, W.A, Astini, R.A. (Eds.), The Conasauga Formation and Equivalent Units in the Appalachian Thrust Belt in Alabama. Alabama Geological Society, Tuscaloosa, AL, pp. 19–40. (Alabama Geological Society 31st Annual Field Trip Guidebook.)

USGS, 2008. Assessment of Undiscovered Oil Resources in the Devonian-Mississippian Bakken Formation, Williston Basin Province, Montana and North Dakota. United States Geological Survey, Reston Virginia (Fact Sheet No. 2008-3021).

Walter, L.M., McIntosh, J.C., Budai, J.M., Martini, A.M., 2000. Hydrogeochemical controls on gas occurrence and production in the New Albany Shale. Gastips 6 (2), 14–20.

Williams, J., Kramer, H., 2011. Montney Shale formation evaluation and reservoir characterization case study well comparing 300 m of Core and Log Data in the upper and lower Montney. Proceedings of the 2011 CSPG CSEG CWLS Convention, Calgary, Alberta, Canada.

Zuber, M.D., Williamson, J.R., Hill, D.G., Sawyer, W.K., Frantz, J.H., 2002. A Comprehensive Reservoir Evaluation of a Shale Gas Reservoir—The New Albany Shale. SPE Paper No. 77469. Proceedings of the Annual Technical Conference and Exhibition, September 20–October 2, San Antonio, TX.

CHAPTER 3

Production Technology

3.1 INTRODUCTION

Recall that shale is a sedimentary rock that is predominantly comprised of very fine-grained clay particles deposited in a thinly laminated texture (see Chapter 1). These formations were originally deposited as mud in low-energy depositional environments, such as tidal flats and swamps, where the clay particles fall out of suspension. During the deposition of these sediments, organic matter is also deposited. Deep burial of this mud results in a layered rock (*shale*), which actually describes the very fine grains and laminar nature of the sediment, not rock composition, which can differ significantly between shale formations.

Shale gas resources are becoming an important energy source for meeting rising energy demands in the next several decades. Development of horizontal drilling and hydraulic fracturing is crucial for economic production of shale gas reservoirs, but it must be performed with caution and as a multidisciplinary approach (King, 2010). Commercial successes in the Barnett Shale, which is currently the largest producing natural gas field, and other shale plays in the United States have made shale gas exploration possible and development has begun to spread all around the world.

Massive hydraulic fractures are created to effectively connect a huge reservoir area to the wellbore when the wellbore is drilled in the direction of minimum horizontal stress. Maximizing the total stimulated reservoir volume (SRV) plays a major role in successful economic gas production (Yu and Sepehrnoori, 2013). Despite the success of shale gas development recently, it is difficult to predict well performance and evaluate economic viability for other shale resources with certainty because high risk and uncertainties are involved.

Briefly, *conventional gas reservoirs* contain *free* gas in interconnected pore spaces that can flow easily to the wellbore, i.e., natural flow is possible (see Chapter 1). On the other hand, *unconventional gas reservoirs* (i.e., shale gas reservoirs) produce from low-permeability (tight and now ultra-tight) formations (see Chapter 1). The gas is often sourced from the reservoir rock itself, adsorbed onto the matrix. Due to the low permeability of these formations, it is necessary to stimulate the reservoir by creating a fracture network to give enough surface area to allow sufficient production from the additional *enhanced* reservoir permeability.

Thus, the uncertainties of reservoir properties and fracture parameters have a significant effect on shale gas production, making the process of optimization of hydraulic fracturing treatment design for economic gas production much more complex. It is extremely important to identify reasonable ranges for these uncertainty parameters and evaluate their effects on well performance, because the detailed reservoir properties for each wellbore are difficult to obtain.

The optimization of critical hydraulic fracture parameters, such as fracture spacing, fracture half-length, and fracture conductivity, which control well performance, is important to obtain the most economical scenario. The cost of hydraulic fracturing of horizontal wells is expensive. Therefore, the development of a method quantifying uncertainties and optimization of shale gas production with economic analysis in an efficient and practical way is clearly desirable (Zhang et al., 2007).

In reality, the ultra-low permeability of shale ranges from 10 to 100 nD (10^{-6} mD), illustrating that shale gas reservoirs are required to be artificially fractured in order to make low-permeability formations produce economically. Typically, the Barnett Shale reservoir exhibits a net thickness of 50−600 ft, porosity of 2−8%, and total organic carbon (TOC) of 1−14% found at depths ranging from 1000 to 13,000 ft (Cipolla et al., 2010). Furthermore, it has been reported that fracture spacing varies in the range from 100 to 700 ft in actual hydraulic fracturing operations in three Barnett Shale wells (Grieser et al., 2009) and well performance of Barnett Shale changes significantly with changing produced fluid type, depth, and formation thickness (Hale and William, 2010). Also, well productivity in the Barnett Shale is highly dependent on the type of completion method implemented and the large hydraulic fracture treatments (Ezisi et al., 2012).

As stated earlier (Chapter 1), shale has very low permeability (measured in nanodarcies). As a result, many wells are required to deplete the reservoir, and special well design and well stimulation techniques are required to deliver production rates of sufficient levels to make a development economic (Schweitzer and Bilgesu, 2009). Horizontal drilling and fracture stimulation have both been crucial in the development of the shale gas industry (Houston et al., 2009).

Natural gas will not readily flow to vertical wells because of the low permeability of shale. This can be overcome by drilling horizontal wells, where the drill bit is steered from its downward trajectory to follow a horizontal trajectory for 1−2 miles, thereby exposing the wellbore to as much of the reservoir as possible.

The use of horizontal drilling in conjunction with hydraulic fracturing has greatly expanded the ability of producers to profitably recover natural gas and oil from low-permeability geologic plays, particularly shale resources (EIA, 2011). Application of fracturing techniques to stimulate oil and gas production began to grow rapidly in the 1950s, although experimentation dates back to the nineteenth century. Starting in the mid-1970s, a partnership of private operators, the US Department of Energy (US DOE) and predecessor agencies, and the Gas Research Institute (GRI), endeavored to develop technologies for the commercial production of natural gas from the relatively shallow Devonian (Huron) Shale in the eastern United States. This partnership helped foster technologies that eventually became crucial to the production of natural gas from shale rock, including horizontal wells, multistage fracturing, and slick-water fracturing. Practical application of horizontal drilling to oil production began in the early 1980s, by which time the advent of improved downhole drilling motors and the invention of other necessary supporting equipment, materials, and technologies (in particular, downhole telemetry equipment) had brought some applications within the realm of commercial viability (EIA, 2011).

All shale gas reservoirs require *fracture stimulation* to connect the natural fracture network to the wellbore (Gale et al., 2007). As drilling is completed, multiple layers of metal casing and cement are placed around the wellbore. After the well is completed, a fluid composed of water, sand, and chemicals is injected under high pressure to crack the shale, increasing the permeability of the rock and easing the flow of natural gas. A portion of the fracturing fluid will return through the well to the surface (*flowback*) due to the subsurface pressures. The volume of fluid will steadily reduce and be replaced by natural gas production.

The fissures created in the fracking process are held open by the sand particles so that natural gas from within the shale can flow up through the well. Once released through the well, the natural gas is captured, stored, and transported to the relevant site processing unit.

Furthermore, each gas shale basin is different and each has a unique set of exploration criteria and operational challenges. The technology was primarily developed in the Texas Barnett Shale and applied to other shale lay resources, often with a one-method-fits-all approach. However, there is now a realization that the Barnett Shale technology needs to be adapted to other shale gas resources in a scientifically technologically structured manner.

While it might be a rule-of-thumb that unconventional resources need unconventional techniques, it is clear that the poorer the reservoir the better the technology and accuracy of data needed to be able to fully characterize and develop each reservoir (resource) (Grieser and Bray, 2007).

In fact, as shale gas resources have emerged as a viable energy source, their characterization using geophysical methods has gained significance (Chopra et al., 2012). The organic content in these shale formations which are measured by the TOC content, influence the compressional and shear velocities as well as the density and anisotropy in these formations. Consequently, detecting changes in the TOC content from the surface seismic response is a necessary step in reservoir characterization. And, in addition to the TOC content, different shale formations have different properties in terms of maturation, gas-in-place, permeability, and brittleness.

The realization is that typical shale reservoirs are more expensive and labor intensive than conventional reservoirs and the expertise needed to characterize reservoir and stimulation treatments is highly specialized.

Prior to recovery proper, a number of vertical wells (perhaps only two or three) are drilled and fractured to determine if shale gas is present and can be extracted. This exploration stage may include an appraisal phase where more wells (perhaps 10—15) are drilled and fractured to: characterize the shale; examine how fractures will tend to propagate; and establish if the shale could produce gas economically.

Further wells may be drilled (perhaps reaching a total of 30) to ascertain the long-term economic viability of the shale.

Once the reservoir properties and contents have been defined, the drilling program and recovery operations will commence.

3.2 WELL DRILLING AND COMPLETION

Natural gas in shale formations will not readily flow to any vertical well drilled through it because of the low permeability of the shale. This can be overcome to some extent by drilling horizontal wells, where, as mentioned above, the drill bit is steered from its downward trajectory to follow a horizontal trajectory for 1 mile or more to thereby exposing the wellbore to as much reservoir as possible. By drilling horizontally, the wellbore may intersect a greater number of naturally existing fractures in the reservoir—the direction of the drill path is chosen based on the known fracture trends in each area. However, some shale formations can only be drilled with vertical wells because of the risk of the borehole collapsing.

Thus, production of gas from shale formations is a multiscale and multimechanism process. Fractures provide the permeability for gas to flow, but contribute little to the overall gas storage capacity. The porosity of the matrix provides most of the storage capacity, but the matrix has very low permeability. Gas flow in the fractures occurs in a different flow regime than for gas flow in the matrix. Because of these differing flow regimes, the modeling of production performance in fractured shale formations is far more complex than for conventional reservoirs, and scaling modeling results up to the field level is very challenging. This in turn makes it difficult to confidently predict production performance and devise optimal depletion strategies for shale resources.

Thus, in order to ensure the optimal development of shale gas resources it is necessary to build a comprehensive understanding of geochemistry, geological history, multiphase flow characteristics, fracture properties (including an understanding of the fracture network), and production behavior across a variety of shale plays. It is also important to develop knowledge that can enable the scaling up of pore-level physics to reservoir-scale performance prediction, and make efforts to improve core analysis techniques to allow accurate determination of the recoverable resource.

For example, unconventional resources require a high well density for full development. Technology that can reduce well costs and increase wellbore contact with the reservoir can make a significant impact on costs, production rates, and ultimate recovery. Multilateral drilling, whereby a number of horizontal sections can be created from a single vertical wellbore, and coiled tubing drilling, to decrease costs, represent potential options for future unconventional gas development.

A combination of steel casing and cement in the well provides an essential barrier to ensure that high-pressure gas or liquids from deeper down cannot escape into shallower rock formations or water aquifers. This barrier has to be designed to withstand the cycles of stress it will endure during the subsequent hydraulic fracturing, without suffering any cracks.

The design aspects that are most important to ensure a leak-free well include the drilling of the wellbore to specification (without additional twists, turns, or cavities), the positioning of the casing in the center of the wellbore before it is cemented in place (this is done with centralizers placed at regular intervals along the casing as it is run in the hole, to keep it away from the rock face), and the correct choice of cement. The cement design needs to be studied both for its liquid properties during pumping (to ensure that it gets to the right place) and for its mechanical strength and flexibility, so that it remains intact. The setting time of the cement is also a critical factor—cement that takes too long to set may have reduced strength; equally, cement that sets before it has been fully pumped into place requires difficult remedial action.

Most shale gas resources are located at depths of 6000 ft or more below ground level, and can be relatively thin (e.g., the Marcellus Shale formation is between 50 and 200 ft thick depending on location). The efficient extraction of gas from such a thin layer of rock requires drilling horizontally through the shale, which is accomplished by drilling vertically downward until the drill bit reaches a distance of around 900 ft from the shale formation. At this point, a directional drill is used to create a gradual 90° curve, so that the *wellbore* becomes horizontal as it reaches optimal depth within the shale. The wellbore then follows the shale formation horizontally for 5000 ft or more. Multiple horizontal wells accessing different parts of the shale formation can be drilled from a single pad. Thus, horizontal drilling reduces the footprint of these operations by enabling a large area of shale to be accessed from a single pad.

Thus, in the process, a large number of fractures are created mechanically in the rock, thus allowing the natural gas and/or crude oil trapped in subsurface formations to move through those fractures to the wellbore from where it can then flow to the surface. Fracking can both increase production rates and increase the total amount of gas that can be recovered from a given volume of shale. Pump pressure causes the rock to fracture, and water carries sand (*proppant*) into the hydraulic fracture to prop it open allowing the flow of gas. Whilst water and sand are the main components of hydraulic fracture fluid, chemical additives are often added in small concentrations to improve fracturing performance.

At various stages in the drilling process, drilling is stopped and steel casing pipe is installed in the wellbore. Cement is pumped into the *annulus* or void space between the casing and the surrounding mineral formation. After the wellbore reaches a depth below the deepest fresh-water aquifer, casing and cement are installed to protect the water from contamination due to the drilling process. Additional casing and cementing along the entire wellbore occurs after the well has reached its full horizontal length. This process is intended to prevent leakage of natural gas from the well to the rock layers between the shale formation and the surface, as well as to prevent the escape of natural gas to the surface through the annulus. The casing surrounding the horizontal section of the well through the shale formation is then per-forated using small explosives to enable the flow of hydraulic fracturing fluids out of the well into the shale and the eventual flow of natural gas out of the shale into the well.

3.2.1 Horizontal Drilling

During the hundred years or so of the existence of the petroleum and natural gas industries, drilling technology has progressed to the point of allowing the driller to turn corners by making the drill bit progress on a horizontal track while accurately staying within a narrow directional and vertical window. Because the horizontal portion is easily controlled, the well is able to drain shale gas resources from a geographical area that is much larger than a single vertical well in the same shale formation.

The primary differences between modern shale gas development and conventional natural gas development are the extensive uses of horizontal drilling and high-volume hydraulic fracturing. The use of horizontal drilling has not introduced any new environmental concerns. In fact, the reduced number of horizontal wells needed coupled with the ability to drill multiple wells from a single pad has significantly reduced surface disturbances and associated impacts to wildlife, dust, noise, and traffic. Where shale gas development has impinged on urban and industrial settings, regulators and industry have developed special practices to alleviate nuisance impacts, impacts to sensitive environmental resources, and interference with existing businesses.

Using the Marcellus Shale resource in Pennsylvania as an example, a vertical well may only drain a cylinder of shale 1320 ft in diameter and as little as 50 ft high. By comparison, a horizontal well may extend from 2000 to 6000 ft in length and drain a volume up to 6000 ft by 1320 ft by 50 ft in thickness, an area about 4000 times greater than that drained by a vertical well. The increase in drainage creates a number of important advantages for horizontal over vertical well concerns, particularly with respect to associated environmental issues.

Thus, horizontal drilling is a technique that allows the wellbore to come into contact with significantly larger areas of hydrocarbon bearing rock than in a vertical well. As a result of this increased contact, production rates and recovery factors can be increased. As the technology for horizontal drilling and fracking has improved, the use of horizontal drilling has increased significantly. An important role that horizontal drilling has played is in development of the natural gas shale resources. These low-permeability rock units contain significant amounts of gas and are present beneath very large parts of North America.

Most horizontal wells begin at the surface as a vertical well. Drilling progresses until the drill bit is a few hundred feet above the target rock unit. At that point, the pipe is pulled from the well and a hydraulic motor is attached between the drill bit and the drill pipe. The hydraulic motor is powered by a flow of drilling mud down the drill pipe. It can rotate the drill bit without rotating the entire length of drill pipe between the bit and the surface. This allows the bit to drill a path that deviates from the orientation of the drill pipe. After the motor is installed, the bit and pipe are lowered back down the well and the bit drills a path that steers the wellbore from vertical to horizontal over a distance of a few hundred feet. Once the well has been steered to the desired angle, straight-ahead drilling resumes and the well follows the target rock unit. Keeping the well in a thin rock unit requires careful navigation. Downhole instruments are used to determine the azimuth and orientation of the drilling. This information is used to steer the drill bit.

The Barnett Shale of Texas, the Fayetteville Shale of Arkansas, the Haynesville Shale of Louisiana and Texas, and the Marcellus Shale of the Appalachian Basin are examples of shale gas resources (Chapter 2). In these rock units, the challenge is to recover gas from very tiny pore spaces in a low-permeability rock unit (Gubelin, 2004). To stimulate the productivity of wells in organic-rich shale, companies drill horizontally through the rock unit and then use hydraulic fracturing to produce artificial permeability. Done together, horizontal drilling and hydraulic fracturing can make a productive well where a vertical well would have produced only a small amount of gas.

In fact, the productive potential of the Haynesville Shale was not fully realized until horizontal drilling and hydrofracturing technologies were demonstrated in other unconventional shale reservoirs. The hydrofracturing process—which is accomplished by sealing off a portion of the well and injecting water or gel under very high pressure into the isolated portion of the hole creating high pressure to create fractures in the rock and open them up—helps liberate gas from the shale and horizontal drilling allows a single well to drain a much larger volume of rock than a traditional vertical well.

In some geological settings, it is more appropriate to directionally drill s-shaped wells from a single pad to minimize surface disturbance. S-shaped wells are drilled vertically several thousand feet, and then extended in arcs beneath the Earth's surface.

During drilling, mobile drilling units are moved between wells on a single pad. This avoids dismantling and reassembling of drilling equipment for each well, making the process quicker and thereby saving resources.

3.2.2 Pad Drilling

When multiple wells are drilled from the same pad, it is often referred to as *pad drilling*—as many as six to eight horizontal wells can originate from the same pad.

Typically, the well pad drains an area that is rectangular (spacing unit, unit or pool), which is usually about one half mile wide by two miles long with the pad itself positioned at the center of the rectangle. The majority of the surface area in the rectangle is not required for the well pad and will be left completely undisturbed. The well pad is generally on the order of 4–5 acres that is cleared, leveled, and surfaced over for siting the drilling rig, trucks, and various other equipment required for drilling and completion activities. The approach allows a drilling company to develop two separate formations on two separate spacing units simultaneously, thereby increasing production efficiency. It also allows the company to recover more of the available resources in a reservoir.

Pad drilling may be accomplished through the use of a movable flex or suitable-for-the-purpose drilling rigs with the intent to drill as many wells on a pad as are economically feasible. Drilling more wells on a pad is considered to help minimize the environmental impact (*environmental footprint*) of the drilling operation.

In shale drilling, it is becoming increasingly common to use a single drill pad to develop as large an area of the subsurface as possible. One surface location may be used to drill multiple wells. Pad drilling increases the operational efficiency of gas production and reduces infrastructure costs and land use. Any negative impact upon the surface environment is therefore mitigated. Such technologies and practices developed by industry serve to reduce environmental impacts from shale gas operations.

3.2.3 Stacked Wells

Drilling stacked horizontal wells may be possible where the shale is sufficiently thick or multiple shale rock strata are found layered on top of each other. One vertical wellbore can be used to produce gas from horizontal wells at different depths. As in pad drilling, the environmental impact on the surface is mitigated as a result of reduced land use. This technology can be particularly beneficial in the thicker shale.

One area where this technology is being employed is in the Pearsall and Eagle Ford plays in southern Texas. Higher efficiency can be achieved as surface facilities are shared. As in pad drilling, the environmental impact on the surface is mitigated as a result of reduced land use.

3.2.4 Multilateral Drilling

Multilateral drilling is similar to stacked drilling in that it involves the drilling of two or more horizontal wells from the same vertical wellbore. With multilateral drilling, the horizontal wells access different areas of the shale at the same depth, but in different directions. Drilling multilateral wells makes it possible for production rates to be increased significantly for a reduced incremental cost.

3.2.5 Well Completion

Once the well has been drilled, the final casing cemented in place across the gas-bearing rock has to be perforated in order to establish communication between the rock and the well (Britt and Smith, 2009; LeCompte et al., 2009; Leonard et al., 2007). The pressure in the well is then lowered so that hydrocarbons can flow from the rock to the well, driven by the pressure differential. With shale and tight gas, the flow will be very low, because of the low permeability of the rock. As the rate of hydrocarbon flow determines directly the cash flow from the well, low flow rates can mean there is insufficient revenue to pay for operating expenses and provide a return on the capital invested. Without additional measures to accelerate the flow of hydrocarbons to the well, the operation is then not economic.

Several technologies have been developed over the years to enhance the flow from low-permeability reservoirs. Acid treatment, involving the injection of small amounts of strong acids into the reservoir to dissolve some of the rock minerals and enhance the permeability of the rock near the wellbore, is probably the oldest and is still widely practiced, particularly in carbonate reservoirs. Wells with long horizontal or lateral sections (horizontal wells) can increase dramatically the contact area between the reservoir rock and the wellbore, and are likewise effective in improving project economics. Hydraulic fracturing, developed initially in the late 1940s, is another effective and commonly practiced technology for low-permeability reservoirs. When rock permeability is extremely low, as in the case of shale gas or light tight oil, it often takes the combination of horizontal wells and hydraulic fracturing to achieve commercial rates of production.

Even though the well casing is perforated, little natural gas will flow freely into the well from the shale. Fracture networks must be created in the shale to allow gas to escape from the pores and natural fractures where it is trapped in the rock. This is accomplished through the process of hydraulic fracturing. In this process, typically several million gallons of a fluid composed of 98−99% w/w water and *proppant* (usually sand) is pumped at high pressure into the well. The rest of the fracking fluid (0.5−2% by volume) is composed of a blend of chemicals, often proprietary, that enhance the fluid's properties. These chemicals typically include acids to clean the shale to improve gas flow; *biocides* to prevent organisms from growing and clogging the shale fractures; corrosion and scale inhibitors to protect the integrity of the well, gels, or gums that add viscosity to the fluid and suspend the proppant; and friction reducers that enhance flow and improve the ability of the fluid to infiltrate and carry the proppant into small fractures in the shale.

This fluid pushes through the *perforations* in the well casing and forces fractures to open in the shale—connecting pores and existing fractures and creating a pathway for natural gas to flow back to the well. The proppant lodges in the fractures and keeps them open once the pressure is reduced and the fluid flows back out of the well. Approximately 1000 ft of wellbore is hydraulically fractured at a time, so each well must be hydraulically fractured in multiple stages, beginning at the furthest end of the wellbore. Cement plugs isolate each hydraulic fracture stage and must be drilled out to enable the flow of natural gas up the well after all hydraulic fracturing is complete.

Once the pressure is released, fluid (commonly referred to as *flowback water*) flows back out through the top of the well. The fluid that is recovered not only contains the proprietary blend of chemicals present in the hydraulic fracturing fluid but may also contain chemicals naturally present in the reservoir, including hydrocarbons, salts, minerals, and naturally occurring radioactive materials that leach into the fluid from the shale or result from mixing of the hydraulic fracturing fluid with brine (e.g., salty water) already present in the formation. The chemical composition of the water produced from the well varies significantly according to the formation and the time after well completion, with early flowback water resembling the hydraulic fracturing fluid but later converging on properties more closely resembling the brine naturally present in the formation.

In many cases, flowback water can be reused in subsequent hydraulic fracturing operations; this depends upon the quality of the flowback water and the economics of other management alternatives. Flowback water that is not reused is managed through disposal. While past disposal options sometimes involved direct dumping into surface waters or deposit at ill-equipped wastewater treatment plants, most disposal now occurs at *Class II injection wells* as regulated by the US Environmental Protection Agency. These injection wells place the flowback water in underground formations isolated from drinking water sources.

3.2.6 Production, Abandonment, and Reclamation

Once wells are connected to processing facilities, the main production phase can begin. During production, wells will produce hydrocarbons and waste streams, which have to be managed. But the well site itself is now less visible: a *Christmas tree* of valves, typically 3–4 ft high, is left on top of the well, with production being piped to processing facilities that usually serve several wells; the rest of the well site can be reclaimed.

In some cases, the operator may decide to repeat the hydraulic fracturing procedure at later times in the life of the producing well, a procedure called refracturing. This was more frequent in vertical wells but is currently relatively rare in horizontal wells, occurring in less than 10% of the horizontal shale gas wells drilled in the United States. The production phase is the longest phase of the lifecycle. For a conventional well, production might last 30 years or more. For an unconventional development, the productive life of a well is expected to be similar, but shale gas wells typically exhibit a burst of initial production and then a steep decline, followed by a long period of relatively low production. Output typically declines by between 50% and 75% in the first year of production, and most recoverable gas is usually extracted after just a few years.

During production, gas that is recovered from the well is sent to small-diameter gathering pipelines that connect to larger pipelines that collect gas from a network of production wells. Because large-scale shale gas production has only been occurring very recently, the production lifetime of shale gas wells is not fully established.

Although there is substantial debate on the issue, it is generally observed that shale gas wells experience quicker production declines than does conventional natural gas production. In the Fayetteville play in north-central Arkansas, it has been estimated that half of a well's lifetime production, or estimated ultimate recovery, occurs within its first 5 years. Once a well no longer produces at an economic rate, the wellhead is removed, the wellbore is filled with cement to prevent leakage of gas into the air, the surface is reclaimed (either to its prewell state or to another condition agreed upon with the landowner), and the site is abandoned to the holder of the land's surface rights.

Like any other well, a shale gas well is abandoned once it reaches the end of the producing life when extraction is no longer economic or possible. As with any gas-producing wells, at the end of their economic life, wells need to be safely abandoned, facilities dismantled and land returned to its natural state or put to new appropriate productive use. Long-term prevention of leaks to aquifers or to the surface is particularly important—sections of the well are filled with cement to prevent gas flowing into water-bearing zones or up to the surface.

Since much of the abandonment will not take place until production has ceased, the regulatory framework needs to ensure that the companies concerned make the necessary financial provisions and maintain technical capacity beyond the economic life of the reservoir to ensure that abandonment is completed satisfactorily, and well integrity maintained over the long term.

3.3 HYDRAULIC FRACTURING

Hydraulic fracturing has been a key technology in making shale gas an affordable addition to the national energy supply, and the technology has proved to be an effective stimulation technique (Arthur et al., 2009; Spellman, 2013). While some challenges exist with water availability and water management (see Chapter 5), innovative regional solutions are emerging that allow shale gas development to continue while ensuring that the water needs of other users are not affected and that surface and ground water quality is protected.

Thus, the first stage in the completion process is then to perforate the well, which refers to explosively punching a hole in the casing or liner, to connect the wellbore to the reservoir. This final stage of the completion process involves running perforating guns (a length of shaped explosive charges) down to the desired depth and firing them to perforate the casing or liner at the required depth.

The second stage is then to hydraulically fracture the well by pumping fluid and proppant at sufficiently high pressures.

Shale is a sedimentary rock that is predominantly comprised of very fine-grained clay particles deposited in a thinly laminated texture, which is fine-grained and laminar and can differ significantly between various shale formations. These rocks were originally deposited as mud in low-energy depositional environments, such as tidal flats and swamps, where the clay particles fall out of suspension. During the deposition of these sediments, organic matter is also deposited, which is measured by the *total organic content* (TOC).

The permeability of typical shale formations (i.e., the ability of fluids to pass through the shale) is very low (in fact, *ultra-low*) compared to conventional oil and gas reservoirs (i.e., nanodarcy (10^{-9} darcy) in shale formations versus millidarcy (10^{-2} darcy) in conventional sandstone formations). In effect, the hydrocarbons are trapped and unable to flow under normal circumstances in shale, and usually only able to migrate out over geologic time. The slow migration of hydrocarbons from shale formations into shallower sandstone reservoirs and carbonate reservoirs has been the source of most conventional oil and gas fields; hence, shale formations have historically been thought of as source and seal rocks, rather than potential reservoirs, but much of the hydrocarbon still remains bound in the shale.

Historically, there has not been any real need or desire to try to develop low-productivity shale reservoirs as they were not economically attractive, though the potentially huge resource has always been suspected. However, recently shale gas development in the United States has been aggressively pursued as was (even in times of plenty) a need to secure low-risk/cost future gas.

Briefly, hydraulic fracturing involves pumping a (fracturing) fluid into a formation at a calculated predetermined rate and pressure to be able to fracture (crack) the shale and create fractures in the formation. Shale gas development typically uses water or water-based fluids as the fracture fluids, mixed with a small amount of various additives.

Sand is the usual *proppant* material and is needed to maintain open fractures once the fluid pumping has been terminated and the fluids have passed into the formation. Initially, fractures were considered to grow at a relatively regular rate and were identical in shape and size at any point in time. However, as knowledge of the fracturing technology has progressed it is now obvious that fracture growth is complicated and involve many more considerations.

Hydraulic fracturing (*fracking*, *fraccing*, or *fracing*) has been widely used for more than 50 years or so by the oil and gas industry to improve low-permeability reservoirs. Fluid (often water, carbon dioxide, nitrogen gas, or propane) is pumped down the well until the pressure surpasses the rock strength and causes the reservoir to crack. The fluid pumped down the well is loaded with proppant— often 100 tons (ca.225,000 lb) or more of ceramic beads or sand— that infiltrates the formation and help to prop the fractures open, which are at risk of closing once the pressure is released. The choice of the fluid used depends on many factors, including whether clay in the reservoir is sensitive to water (some clays swell in the presence of freshwater) or whether the reservoir happens to respond better to particular fluids, this usually only being determined through experimentation.

Two factors increase the ability of shale to fracture: (1) the presence of *hard minerals* and (2) the *internal pressure* of the shale.

The presence of hard minerals such as silica (and to a lesser extent calcite), which break like glass, induce fractures in the shale when under pressure. Clay, however, tends to absorb more of the pressure and often bends under applied hydraulic pressure without breaking. Therefore, silica-rich shale formations are good candidates for fracking. In terms of the internal pressure of the shale, over-pressured shale formations develop during the generation of natural gas— because of the low permeability, much of the gas cannot escape and builds in place, increasing the internal pressure of the rock. Therefore, the artificially created fracture network can penetrate further into the formation because the shale is already closer to the breaking point than in normally pressured shale formations.

3.3.1 General Aspects

Hydraulic fracturing is one of the key drivers to shale gas development because of the low to ultra-low permeability factor involved. Also key to shale gas development is the presence of natural fractures and planes of weakness that can result in complex fracture geometries during stimulation (Reddy and Nair, 2012). Furthermore, the presence and ability to open and maintain flow in both primary and secondary natural fracture systems are critical to shale gas production (King, 2010).

Hydraulic fracturing is a technology that involves pumping water, sand, and a small amount of chemical additives into the well to fracture the rock, freeing the natural gas. This is common in oil and natural gas development—the technology has been used since the 1940s in more than one million wells in the United States. In fact, 90% of oil and gas wells in the United States undergo hydraulic fracturing to enhance production flow rates.

The development of large-scale shale gas production is changing the US energy market, generating expanded interest in the usage of natural gas in sectors such as electricity generation and transportation. At the same time, there is much uncertainty of the environmental implications of hydraulic fracturing and the rapid expansion of natural gas production from shale formations.

Water for fracturing can come from surface water sources (such as rivers, lakes, or the sea), or from local boreholes (which may draw from shallow or deep aquifers and which may already have been drilled to support production operations), or from further afield (which generally requires trucking). Transportation of water from its source and to disposal locations can be a large-scale activity.

In areas of water scarcity, the extraction of water for drilling and hydraulic fracturing (or even the production of water, in the case of coalbed methane) can have broad and serious environmental effects. It can lower the water table, affect biodiversity, and harm the local ecosystem. It can also reduce the availability of water for use by local communities and in other productive activities, such as agriculture.

Limited availability of water for hydraulic fracturing could become a significant constraint on the development of tight gas and shale gas in some water-stressed areas. In China, for example, the Tarim Basin in the Xinjiang Uyghur Autonomous Region holds some of the country's largest shale gas deposits, but also suffers from severe water scarcity. Although not on the same scale, in terms of either resource endowment or water stress, a number of other prospective deposits occur in regions that are already experiencing intense competition for water resources. The development of China's shale gas industry has to date focused on the Sichuan Basin, in part, because water is much more abundant in this region.

Hydraulic fracturing dominates the freshwater requirements for unconventional gas wells and the dominant choice of fracturing fluid for shale gas, "slick-water," which is often available at the lowest cost and which in some shale reservoirs may also bring some gas production benefits, is actually the most demanding in terms of water needs. Much attention has accordingly been given to approaches which might reduce the amount of water used in fracturing. Total pumped volumes (and therefore water volumes required) can be decreased through the use of more traditional, high viscosity, fracturing fluids (using polymers or surfactants), but these require a complex cocktail of chemicals to be added.

Foamed fluids, in which water is foamed with nitrogen or carbon dioxide, with the help of surfactants (as used in dishwashing liquids), can be attractive as 90% of the fluid can be gas and this fluid has very good proppant-carrying properties. Water can, indeed, be eliminated altogether by using hydrocarbon-based fracturing fluids, such as propane or gelled hydrocarbons, but their flammability makes them more difficult to handle safely at the well site. The percentage of fracturing fluid that gets back-produced during the flowback phase varies with the type of fluid used (and the shale characteristics), so the optimum choice of fluid will depend on many factors: the availability of water, whether water recycling is included in the project, the properties of the shale reservoir being tapped, the desire to reduce the usage of chemicals, and the economics (Blauch et al., 2009).

Unlike conventional mineral formations containing natural gas deposits, shale has low permeability, which naturally limits the flow of gas or water. In shale formations, natural gas is held in largely unconnected pores and natural fractures. Hydraulic fracturing is the method commonly used to connect these pores and allow the gas to flow. The process of producing natural gas from shale deposits involves many steps in addition to hydraulic fracturing, all of which involve potential environmental impacts. Hydraulic fracturing is often misused as an umbrella term to include all of the steps involved in shale gas production. These steps include road and well pad construction, drilling the well, *casing*, perforating, hydraulic fracturing, *completion*, production, abandonment, and reclamation.

A common issue encountered in hydraulic fracturing operations in gas shale formations is the variability and unpredictability of the outcome of hydraulic fracturing. Industry experiences show that injection pressures required to fracture the formation (fracture gradient) oftentimes vary significantly along a well, and there can be intervals where the formation cannot be fractured successfully by fluid injection. The use of real-time fracture mapping allows for on-the-fly changes in fracture design. Mapping also impacts the perforation strategy and restimulation designs to maximize the *effective stimulation volume* (the reservoir volume that has been effectively contacted by the stimulation treatment as determined by microseismic event locations and density). A correlation of microseismic activity with log data allows estimation of fracture geometry to be made after which the data can be used to design a stimulation that has the greatest chance of maximizing production (Baihly et al., 2006; Daniels et al., 2007; Fisher et al., 2004).

Shale gas reservoirs also respond to fluid injection in a variety of modes. As observed through microseismic monitoring, distribution of activated seismicity can be confined along a macroscopic fracture plane, but mostly they are dispersed throughout a wide region in the reservoir reflecting the development of a complex fracture network (Cipolla et al., 2009; Das and Zoback, 2011; Maxwell, 2011; Waters et al., 2006).

In recent years, various attempts have been made to optimize the design of transverse fractures of horizontal wells for shale gas reservoirs (Bhattacharya and Nikolaou, 2011; Britt and Smith, 2009; Gorucu and Ertekin, 2011; Marongiu-Porcu et al., 2009; Meyer et al., 2010). In most cases, the optimum design is identified by local sensitivity analysis and usually one variable is varied while keeping all other variables fixed. However, these optimization methods may not provide sufficient insight for screening insignificant parameters and for considering parameter interactions to obtain the optimal design. Hence, the optimization of hydraulic fracturing treatment design for shale gas production remains a challenge.

An additional factor to consider is the shale's thickness. The substantial thickness of shale is one of the primary reasons, along with a large surface area of fine-grained sediment and organic matter for adsorption of gas, that shale resource evaluations yield such high values for TOC content and potential gas producibility. Not surprisingly, a general rule-of-thumb is that thicker shale is a better target. Shale targets such as the Bakken oil play in the Williston Basin (itself a hybrid conventional–unconventional resource), however, are less than 150 ft thick in many areas and are yielding apparently economic rates of gas flow and recovery. The required thickness to economically develop a shale gas target may decrease as drilling and completion techniques improve, as porosity and permeability detection techniques progress in unconventional targets and, perhaps, as the price of gas increases. Such a situation would add a substantial amount of resources and reserves to the shale gas formation.

3.3.2 Fracturing Fluids

Initially, the fluid that is injected does not contain any propping agent (called pad), and creates a fracture that is multidirectional and spreads up, out, and down. The pad creates a fracture that is wide enough to begin accepting a propping agent material. The pad is then followed by the proppant slurry—a mix of the carrier fluid and proppant material. The purpose of the propping agent is to *prop open* the fracture once the pumping operation ceases and the fracture closes.

In deep reservoirs, man-made beads are sometimes used to prop open the fractures but in shallow reservoirs, sand can be used and remains the most common proppant. Once the fracture has been initiated, fluid is continually pumped into the wellbore to extend the created fracture and develop a fracture network. Each formation has different properties and *in situ* stresses so that each hydraulic fracture job is unique and different and is specifically designed for that well by a hydraulic fracturing specialist. The process of designing hydraulic fracture treatments involves identifying properties of the target formation including estimating fracture treating pressure, amount of material, and the desired length for optimal economics.

The fracturing fluid should have a number of properties that are tailored to and optimized for each formation, that is, the fluid should (i) be compatible with the formation rock, (ii) be compatible with the formation fluid, (iii) generate sufficient pressure drop down the fracture to create a wide fracture, (iv) have sufficient lower viscosity to allow clean-up after the treatment, and (v) be cost-effective. Water-based fluids are commonly used and *slick-water* is the most common fluid used for shale gas fracturing, where the major chemical added is a surfactant polymer to reduce the surface tension or friction, so that water can be pumped at low-treating pressures. Other fluids that have been used are oil-based fluids, energized fluids, foams, and emulsions.

Environmental concerns have focused on the fluid used for hydraulic fracturing and the risk of water contamination through leaks of this fluid into ground water. Water itself, together with sand or ceramic beads (the "proppant"), makes up over 99% of a typical fracturing fluid, but a mixture of chemical additives is also used to give the fluid the properties that are needed for fracturing. These properties vary according to the type of formation. Additives (not all of which would be used in all fracturing fluids) typically help to accomplish four tasks:

1. To keep the proppant suspended in the fluid by gelling the fluid while it is being pumped into the well and to ensure that the proppant ends up in the fractures being created. Without this effect, the heavier proppant particles would tend to be distributed unevenly in the fluid under the influence of gravity and would, therefore, be less effective. Gelling polymers, such as guar or cellulose (similar to those used in food and cosmetics) are used at a concentration of about 1%. Cross-linking agents, such as borates or metallic salts, are also commonly used at very low concentration to form a stronger gel. They can be toxic at high concentrations, though they are often found at low natural concentrations in mineral water.

2. To change the properties of the fluid over time. Characteristics that are needed to deliver the proppant deep into subsurface cracks are not desirable at other stages in the process, so there are additives that give time-dependent properties to the fluid, for example to make the fluid less viscous after fracturing, so that the hydrocarbons flow more easily along the fractures to the well. Typically, small concentrations of chelants (such as those used to descale kettles) are used, as are small concentrations of oxidants or

enzymes (used in a range of industrial processes) to break down the gelling polymer at the end of the process.
3. To reduce friction and therefore reduce the power required to inject the fluid into the well. A typical drag-reducing polymer is polyacrylamide (widely used, e.g., as an absorbent in baby diapers).
4. To reduce the risk that naturally occurring bacteria in the water affect the performance of the fracturing fluid or proliferate in the reservoir, producing hydrogen sulfide; this is often achieved by using a disinfectant (biocide), similar to those commonly used in hospitals or cleaning supplies.

Until recently, the chemical composition of fracturing fluids was considered a trade secret and had not been made public. This position has fallen increasingly out of step with public insistence that the community has the right to know what is being injected into the ground. Since 2010, voluntary disclosure has become the norm in most of the United States. The industry is also looking at ways to achieve the desired results without using potentially harmful chemicals. "Slickwater," made up of water, proppant, simple drag-reducing polymers, and biocide, has become increasingly popular as a fracturing fluid in the United States, though it needs to be pumped at high rates and can carry only very fine proppant. Attention is also being focused on reducing accidental surface spills, which most experts regard as a more significant risk of contamination to ground water.

Finally, because of the properties of the shale formation (above), such as (i) the presence of *hard minerals* and (ii) the *internal pressure* of the shale, it may be possible to isolate sections along the horizontal portion of the well, segments of the borehole for one-at-a-time fracking (multistage fraccing). By monitoring the process at the surface and in neighboring wells, it can be determined how far, how extensively, and in what directions the shale has cracked from the induced pressure.

Finally, shale formations can be refracked years later, after production has declined, which may allow (i) the well to access larger areas of the reservoir that may have been missed during the initial hydraulic fracturing or (ii) the reopening of fractures that may have closed due to the decrease in pressure as the reservoir was drained. Even with hydraulic fracturing, wells drilled into low-permeability reservoirs have difficulty in communicating far into the formation. As a result, additional wells must be drilled to access as much gas as possible, employing typically three or four, but up to eight, horizontal wells per section.

3.3.3 Fracturing Fluid Additives

Possible additives for fracturing fluids are chosen according to the task at hand—that is, the properties of the reservoir. These additives include (i) polymers, which allow for an increase in the viscosity of the fluid, together with cross-linkers, (ii) cross-linkers, which increase the viscosity of the linear polymer base gel, (iii) breakers, which are used to break the polymers and cross-link sites at formation temperature, for better clean-up, (iv) biocides, which are used to kill bacteria in the mix water, (v) buffers, which are used to control the pH, (vi) fluid loss additives, which are used to control excessive fluid leak-off into the formation, and (vii) stabilizers, which are used to keep the fluid viscous at higher temperature.

However, it must be emphasized that additives are used for every site and in general as few additives as possible are added to avoid potential environmental contamination (use of the additives must be controlled) and production problems with the reservoir.

A recent innovation in completion technology has been the addition of 3% v/v hydrochloric acid to induced fracturing in the Barnett Shale, which appears to increase the daily flow rate by enhancing matrix permeability and may add to the estimated ultimate recovery (Grieser et al., 2007). In addition, refracturing the reservoir is an option that is becoming more and more commonplace (Cramer, 2008) and can yield additional recoverable reserves.

Rocks with interlaminated shale and siltstone constitute a shale gas target (e.g., Lewis Shale, New Mexico; Colorado Group, Alberta) that may require new techniques for detection in well logs, as well as new completion and drilling techniques. The silt laminations are too thin to be detected on well logs and to allow an accurate determination of how many laminations are present in a given interval. Also, well logs are unable to accurately determine the percentage of porosity in shale or laminations, the degree of water saturation in a reservoir, or the relative degree of permeability in each lamination. Laminations both store gas (free gas) and are pathways of transport for diffusion of gas from shale to the wellbore (Beaton et al., 2009; Pawlowicz et al., 2009; Rokosh et al., 2009).

Laminations are also particularly difficult completion targets. Normally, induced fractures are meant to extend laterally rather than vertically in a reservoir, yet the laminations may span tens of hundreds of feet vertically. Therefore, a horizontal fracture may miss many productive shale and silt laminations. Induced fracturing techniques may have to be altered or new techniques developed for this type of shale gas reservoir.

3.3.4 Fracture Diagnostics

Fracture diagnostics are the techniques used to analyze the created fractures and involve analyzing data before (prefracture analysis), during (real time) and after (postfracture) hydraulic fracture treatment (Barree et al., 2002; Vulgamore et al., 2007). The *raison d'être* for the determination of the dimensions of the created fractures is to determine whether or not the fractures are effectively maintained in an open mode (*propped*) fracture. The diagnostic techniques are generally subdivided into three groups: (1) direct far-field techniques, (2) direct near-wellbore techniques, and (3) indirect fracture techniques.

3.3.4.1 Direct Far-Field Techniques

The *direct far-field* techniques comprise tiltmeter (an instrument designed to measure very small changes from the horizontal level, either on the ground or in subterranean structures) and microseismic fracture mapping techniques, which require delicate instrumentation to be placed in boreholes surrounding and near the well to be fracture treated. Microseismic fracture mapping typically relies on using a downhole receiver array of accelerometers, or geophones, to locate *micro-earthquakes* that are triggered by shear slippage in natural fractures surrounding the hydraulic fracture. As with all monitoring and data collection techniques, however, examination of wells that are typically considered marginal wells is often not justified until the resource has been proved. If the technology is used at the beginning of the development of a field, however, the data and knowledge gained may be worthwhile, and effective development of the shale resource is warranted.

3.3.4.2 Direct Near-Wellbore Techniques

Direct near-wellbore techniques—which consist of tracer logs, temperature logging, production logging, borehole image logging, downhole video logging, and caliper logging—are used in the well that is being fractured to locate the portion of fracture that is very near the wellbore.

However, in shale gas reservoirs, where multiple fractures are likely to exist, the reliability of these direct near-wellbore techniques may be questionable. As such, very few of these direct near-wellbore techniques are used on a routinely-without-question basis to evaluate hydraulic fracture patterns, and if deployed the *direct near-wellbore techniques* are typically used in conjunction with other more reliable techniques.

3.3.4.3 Indirect Fracture Techniques

Indirect fracture techniques consist of hydraulic fracture modeling and matching of the net surface treating pressures, together with subsequent pressure transient test analyses and production data analyses. As fracture treatment data and the postfracture production data are normally available on every well, indirect fracture diagnostic techniques are the most widely used methods to determine the shape and dimensions of both the created and the propped hydraulic fracture.

3.4 PRODUCTION TRENDS

Economic natural gas production from unconventional shale gas reservoirs is achieved by the combination of horizontal drilling and reservoir stimulation by multistage slick-water fracturing. Ideally, every fracture treatment at every stage of the well is successful, but experience from the Barnett Shale and other shale gas reservoirs has shown that not all stages are stimulated equally. The regions of SRV may be different between stages in size and shape, sometimes confined along a plane, or sometimes dispersed widely in the reservoir (Maxwell, 2011; Waters et al., 2006).

Operators also have observed that fracturing pressure (fracture gradient) can vary between stages, sometimes to a point where pump pressures cannot reach the required fracturing pressure to propagate a fracture (Daniels et al., 2007). These variations in the outcome of hydraulic fracturing are caused by the heterogeneity in the rock mechanical/deformation properties, presence of natural fractures, and/or the variations in *in situ* stress within the reservoir.

However, the increasing participation of major oil companies in North American shale gas exploitation has positive implications for the use of best practices and technologies in drilling and processing. Continued development of shale gas in North America and other countries with significant resources will have an impact on the global gas markets; however, this impact is expected to remain moderate in the short to medium term, nothing comparable to what happened in the United States.

The increasing use of shale gas will primarily impact power generation, transport fuels, and the petrochemical industry. In fact, global estimates of proven reserves of shale gas are increasing and will continue to do so as exploration continues. Furthermore, exploitation of shale gas is keeping natural gas prices low, particularly in North America.

3.4.1 Technology

Modern shale gas development is a technologically driven process for the production of natural gas resources. Currently, the drilling and completion of shale gas wells includes both vertical and horizontal wells. In both kinds of wells, casing and cement are installed to protect fresh and treatable water aquifers. The emerging shale gas basins are expected to follow a trend similar to the Barnett Shale play with increasing numbers of horizontal wells as the plays mature. Shale gas operators are increasingly relying on horizontal well completions to optimize recovery and well economics. Horizontal drilling provides more exposure to a formation than does a vertical well.

This increase in reservoir exposure creates a number of advantages over vertical well drilling. Six to eight horizontal wells drilled from only one well pad can access the same reservoir volume as sixteen vertical wells. Using multiwell pads can also significantly reduce the overall number of well pads, access roads, pipeline routes, and production facilities required, thus minimizing habitat disturbance, impacts to the public, and the overall environmental footprint (see Chapter 5).

The other technological key to the economic recovery of shale gas is hydraulic fracturing, which involves the pumping of a fracturing fluid under high pressure into a shale formation to generate fractures or cracks in the target rock formation. This allows the natural gas to flow out of the shale to the well in economic quantities. Ground water is protected during the shale gas fracturing process by a combination of the casing and cement that is installed when the well is drilled and the thousands of feet of rock between the fracture zone and any fresh or treatable aquifers. For shale gas development, fracture fluids are primarily water-based fluids mixed with additives that help the water to carry sand proppant into the fractures. Water and sand make up over 98% of the fracture fluid, with the rest consisting of various chemical additives that improve the effectiveness of the fracture job. Each hydraulic fracture treatment is a highly controlled process that must be designed to the specific conditions of the target formation.

A combination of improved technology and shale-specific experience has also led to improvements in recovery factors and reductions in decline rates. Each shale resource requires its own specific completion techniques, which can be determined through careful analysis of rock properties. The correct selection of well orientation, stimulation equipment, fracture size, and fracking fluids can all affect the performance of a well.

The initial production rate from a particular well is highly dependent on the quality of the fracture and the well completion. In the United States, it has been seen that initial production rates for additional wells have been augmented over time as the resource matures. Initial production rates can be increased by several techniques, in particular by increasing the number of fracture stages and increasing the number of perforations per fracture stage. The quality of the fracture is also improved as fluid properties are developed. Microseismic data can also be used to improve the efficiency of the fracking process.

The primary differences between modern shale gas development and conventional natural gas development are the extensive uses of horizontal drilling and high-volume hydraulic fracturing. The use of horizontal drilling has not introduced any new environmental concerns. In fact, the reduced number of horizontal wells needed coupled with the ability to drill multiple wells from a single pad has significantly reduced surface disturbances and associated impacts on wildlife, dust, noise, and traffic. Where shale gas development has intersected with urban and industrial settings, regulators and industry have developed special practices to alleviate nuisance impacts, impacts to sensitive environmental resources, and interference with existing businesses.

Hydraulic fracturing has been a key technology in making shale gas an affordable addition to the US energy supply, and the technology has proved to be an effective stimulation technique. While some challenges exist with water availability and water management, innovative regional solutions are emerging that allow shale gas development to continue while ensuring that the water needs of other users are not affected and that surface and ground water quality is protected. Taken together, state and federal requirements along with the technologies and practices developed by industry serve to reduce environmental impacts from shale gas operations.

3.4.2 The Future

Development of shale gas resources in Western Europe, including Scandinavia, and Poland has the potential to cut the heavy dependence by western European countries on Russian gas, unless, of course, Russian gas companies gain control of these resources. Furthermore, shale gas discoveries in South America have the potential of realigning the energy relationships on the continent. Argentina, Brazil, and Chile are likely beneficiaries decreasing their dependence on Bolivian gas.

Notwithstanding the environmental moratoriums in some countries, shale gas will be used in many parts of the world. This will place downward pressure on prices of natural gas and lower natural gas prices may lead to significant shifting in power generation and transport fuels.

Finally, shale gas liquids are having a significant impact on the petrochemical industry in North America, which has spillover effects in Europe, the Middle East, and Asia. Further, these liquids are making shale gas more profitable than traditional dry gas reservoirs.

Despite the availability of proven production technologies, environmental impacts are still being queried; in particular the impact is on ground water resources and the possible methane releases associated with current production techniques. These issues are the subject of intense scrutiny at the moment. The supply and use of shale gas is already showing an impact on the fossil fuels energy sector and is not restricted to the global natural gas pricing outlook but its development has become entwined with the global energy mix and emerging nexus in energy—climate—water and its impact on the global energy supplies environment.

Finally, property and mineral rights differ across the world. In the United States, individuals can own the mineral rights for the land they own. In many parts of Asia, Europe, and South America this is not the case. Therefore, unresolved legal issues remain obstacles for shale gas exploitation in many countries across the globe. Given the investment requirements needed to develop shale basins that presently have little to no infrastructure, the legal issues are important to attract the investment needed for exploration and exploitation.

In summary, shale gas reservoirs have heterogeneous geological and geomechanical characteristics that pose challenges to accurate prediction of the response to hydraulic fracturing. Experience in shale gas formations shows that stimulation often results in formation of a complex fracture structure, rather than the planar fracture aligned with the maximum principal stress. The fracture complexity arises from intact rock and rock mass textural characteristics and the *in situ* stress and their interaction with applied loads. Open and mineralized joints and interfaces, and contact between rock units, play an important role in fracture network complexity, which affects the rock mass permeability and its evolution with time.

Currently, the mechanisms that generate these fracture systems are not completely understood, and can generally be attributed to lack of *in situ* stress contrast, rock brittleness, shear reactivation of mineralized fractures, and textural heterogeneity. This clearly indicates the importance of linking the mineralogy, rock mechanics, and geomechanics to determine the future of an unconventional shale resource.

REFERENCES

Arthur, J.D., Bohm, B., Coughlin, B.J., Layne, M., 2009. Evaluating implications of hydraulic fracturing in shale gas reservoirs. Paper No. SPE 121038. Proceedings of the SPE Americas Environmental and Safety Conference, 23–25 March, San Antonio, TX.

Baihly, J., Laursen, P., Ogrin, J., Le Calvez, J.H., Villarreal, R., Tanner, K., et al., 2006. Using microseismic monitoring and advanced stimulation technology to understand fracture geometry and eliminate screenout problems in the Bossier Sand of East Texas. Paper No. SPE 102493. Proceedings of the SPE Annual Conference and Exhibition, 24–27 September, San Antonio, TX.

Barree, R.D., Fisher, M.K., Woodrood, R.A., 2002. A practical guide to hydraulic fracture diagnostics technologies. Paper No. SPE 77442. Proceedings of the SPE Annual Technical Conference and Exhibition, 29 September –2 October 2, San Antonio, TX.

Beaton, A.P., Pawlowicz, J.G., Anderson, S.D.A., Rokosh, C.D., 2009. Rock Eval™ Total Organic Carbon, Adsorption Isotherms and Organic Petrography of the Colorado Group: Shale Gas Data Release. Energy Resources Conservation Board, Calgary, Alberta, Canada (Open File Report No. ERCB/AGS 2008–11).

Bhattacharya, S., Nikolaou, M., 2011. Optimal fracture spacing and stimulation design for horizontal wells in unconventional gas reservoirs. Paper No. SPE 147622. Proceedings of the SPE Annual Technical Conference and Exhibition, 30 October –2 November 2, Denver, CO.

Blauch, M.E., Myers, R.R., Moore, T.R., Houston, N.A., 2009. Marcellus Shale post-frac flow-back waters—where is all the salt coming from and what are the implications. Paper No. SPE 125740. Proceedings of the SPE Regional Meeting, 23–25 September, Charleston, WV.

Britt, L.K., Smith, M.B., 2009. Horizontal well completion, stimulation optimization, and risk mitigation. Paper No. SPE 125526. Proceedings of the SPE Eastern Regional Meeting, 23–25 September, Charleston, WV.

Chopra, S., Sharma, R.K., Keay, J., Marfurt, K.J., 2012. Shale gas reservoir characterization workflows. Proceedings of the SEG Annual Meeting, Las Vegas, Nevada. Society of Exploration Geophysicists, Tulsa, OK.

Cipolla, C.L., Lolon, E.P., Mayerhofer, M.J., Warpinski, N.R., 2009. Fracture design considerations in horizontal wells drilled in unconventional gas reservoirs. Paper No. SPE 119366. Proceedings of the SPE Hydraulic Fracturing Technology Conference, 19–21 January, The Woodlands, TX.

Cipolla, C.L., Lolon, E.P., Erdle, J.C., Rubin, B., 2010. Reservoir modeling in Shale-Gas reservoirs (Paper No. SPE 125530). SPE Res Eval. Eng. 13 (4), 638–653.

Cramer, D.D., 2008. Stimulating unconventional reservoirs: lessons learned, successful practices, areas for improvement. SPE Paper No. 114172. 2008 Unconventional Gas Conference, 10–12 February, Keystone, CO.

Daniels, J., Waters, G., LeCalvez, J., Lassek, J., Bentley, D., 2007. Contacting more of the Barnett Shale through an integration of real-time microseismic monitoring, petrophysics and hydraulic fracture design. SPE Paper No. 110562. Proceedings of the SPE Annual Technical Conference and Exhibition, Anaheim, CA.

Das, I, Zoback, M.D., 2011. Long-Period, Long-Duration seismic events during hydraulic fracture stimulation of a shale gas reservoir. Leading Edge 30, 778–786.

EIA, 2011. Review of Emerging Resources: US Shale Gas and Shale Oil Plays. Energy Information Administration. United States Department of Energy, Washington, DC, July.

Ezisi, L.B., Hale, B.W., William, M., Watson, M.C., Heinze, L., 2012. Assessment of probabilistic parameters for Barnett shale recoverable volumes. Paper No. SPE 162915. Proceedings of the SPE Hydrocarbon, Economics, and Evaluation Symposium, 24–25 September, Calgary, Canada.

Fisher, M.K., Heinze, J.R., Harris, C.D., McDavidson, B.M., Wright, C.A., Dunn, K.P., 2004. Optimizing horizontal completion techniques in the Barnett shale using microseismic fracture mapping. Paper No. SPE 90051. Proceedings of the SPE Annual Technical Conference and Exhibition, 26–29 September, Houston, TX.

Gale, J.F.W., Reed, R.M., Holder, J., 2007. Natural fractures in the Barnett Shale and their importance for hydraulic fracture treatments. Am. Assoc. Pet. Geol. Bull. 91, 603–622.

Gorucu, S.E., Ertekin, T., 2011. Optimization of the design of transverse hydraulic fractures in horizontal wells placed in dual porosity tight gas reservoirs. Paper No. SPE 142040. Proceedings of the SPE Middle East Unconventional Gas Conference and Exhibition, 31 January–2 February, Muscat, Oman.

Grieser, B., Bray, J., 2007. Identification of production potential in unconventional reservoirs. Paper No. SPE 106623. Proceedings of the SPE Production and Operations Symposium, 31 March–3 April, Oklahoma City, OK.

Grieser, B., Wheaton, B., Magness, B. Blauch, M., Loghry, R. 2007. Surface reactive fluid's effect on shale. SPE Paper No. 106815. Proceedings of the SPE Production and Operations Symposium, 31 March–3 April, Society of Petroleum Engineers, Oklahoma City, OK.

Grieser, B., Shelley, B., Soliman, M., 2009. Predicting production outcome from multi-stage, horizontal Barnett completions. Paper No. SPE 120271. Proceedings of the SPE Production and Operation Symposium, 4–8 April, Oklahoma City, OK.

Gubelin, G., 2004. Improving gas recovery factor in the Barnett Shale through the application of reservoir characterization and simulation answers. Proceedings of the Gas Shales: Production and Potential, 29–30 July, Denver, CO.

Hale, B.W., William, M., 2010. Barnett Shale: a resource play—locally random and regionally complex. Paper No. SPE 138987. Proceedings of the SPE Eastern Regional Meeting, 12–14 October, Morgantown, WV.

Houston, N., Blauch, M., Weaver, III, D., Miller, D.S., O'Hara, D., 2009. Fracture-stimulation in the Marcellus Shale—lessons learned in fluid selection and execution. Paper No. SPE 125987. Proceedings of the SPE Regional Meeting, 23–25 September, Charleston, WV.

King, G.E., 2010. Thirty years of gas shale fracturing: what have we learned? Paper No. SPE 133456. Proceedings of the SPE Annual Technical Conference and Exhibition, September, Florence, Italy.

LeCompte, B., Franquet, J.A., Jacobi, D., 2009. Evaluation of Haynesville Shale vertical well completions with a mineralogy based approach to reservoir geomechanics. Paper No. SPE 124227. Proceedings of the SPE Annual Technical Meeting, 4–7 October, New Orleans, LO.

Leonard, R., Woodroof, R.A., Bullard, K., Middlebrook, M., Wilson, R., 2007. Barnett Shale completions: a method for assessing new completion strategies. Paper No. SPE 110809. Proceedings of the SPE Annual Technical Conference and Exhibition, 11–14 November, Anaheim, CA.

Marongiu-Porcu, M., Wang, X., Economides, M.J., 2009. Delineation of application: physical and economic optimization of fractured gas wells. Paper No. SPE 120114. Proceedings of the SPE Production and Operations Symposium, 4–8 April, Oklahoma City, OK.

Maxwell, S., 2011. Microseismic hydraulic fracture imaging: the path toward optimizing shale gas production. Leading Edge 30, 340–346.

Meyer, B.R., Bazan, L.W., Jacot, R.H., Lattibeaudiere, M.G., 2010. Optimization of multiple transverse hydraulic fractures in horizontal wellbores. Paper No. SPE 131732. Proceedings of the SPE Unconventional Gas Conference, 23–25 February, Pittsburgh, PA.

Pawlowicz, J.G., Anderson, S.D.A., Rokosh, C.D., Beaton, A.P., 2009. Mineralogy, Permeametry, Mercury Porosimetry and Scanning Electron Microscope Imaging of the Colorado Group: Shale Gas Data Release. Energy Resources Conservation Board, Calgary, Alberta, Canada (Open File Report No. ERCB/AGS Report 2008-14).

Reddy, T.R., Nair, R.R., 2012. Fracture characterization of shale gas reservoir using connected-cluster DFN simulation. Proceedings of the Second International Conference on Drilling Technology 2012 (ICDT-2012) and First National Symposium on Petroleum Science and Engineering 2012 (NSPSE-2012). Sharma, R., Sundaravadivelu, R., Bhattacharyya, S.K., Subramanian, S.P. (Eds.), 6–8 December, pp. 133–136.

Rokosh, C.D., Pawlowicz, J.G., Berhane, H., Anderson, S.D.A., Beaton, A.P., 2009. Geochemical and Sedimentological Investigation of the Colorado Group for Shale Gas Potential: Initial Results. Energy Resources Conservation Board, Calgary, Alberta, Canada (Open File Report No. ERCB/AGS 2008-09).

Schweitzer, R., Bilgesu, H.I., 2009. The role of economics on well and fracture design completions of Marcellus Shale wells. Paper No. SPE 125975. Proceedings of the SPE Eastern Regional Meeting, 23–25 September, Charleston, WV.

Spellman, F.R., 2013. Environmental Impacts of Hydraulic Fracturing. CRC Press, Taylor & Francis Group, Boca Raton, FL.

Vulgamore, T., Clawson, T., Pope, C., Wolhart, S, Mayerhofer, M., Machovoe, S., et al., 2007. Applying hydraulic fracture diagnostics to optimize stimulations in the Woodford shale. Paper No. SPE 110029. Proceedings of the SPE Annual Technical Conference and Exhibition, 11–14 November, Anaheim, CA.

Waters, G., Heinze, J., Jackson, R., Ketter, A., Daniels, J., Bentley, D., 2006. Use of horizontal well image tools to optimize Barnett shale reservoir exploitation. SPE Paper No. 103202. Proceedings of the SPE Annual Technical Conference and Exhibition, San Antonio, TX.

Yu, W., Sepehrnoori, K., 2013. Optimization of multiple hydraulically fractured horizontal wells in unconventional gas reservoirs. Paper No. SPE 164509. Proceedings of the SPE Production and Operations Symposium, 23–26 March, Oklahoma City, OK.

Zhang, J., Delshad, M., Sepehrnoori, K., 2007. Development of a framework for optimization of reservoir simulation studies. J. Pet. Sci. Eng. 59, 135–146.

CHAPTER 4

Shale Gas Properties and Processing

4.1 INTRODUCTION

Natural gas production from shale gas reservoirs is now proven to be feasible from numerous operations in various shale gas reservoirs in North America, but many challenges still remain in the full exploitation of these unconventional reservoirs. Production from shale gas reservoirs requires stimulation by hydraulic fracturing due to the extremely low permeability of the reservoir rocks. Maximization of reservoir producibility can only be achieved by a thorough understanding of the occurrence and properties of the shale gas resources (see Chapters 1 and 2) as well as the producibility of the gas from the reservoir (see Chapter 3) (Kundert and Mullen, 2009). Although distinct in focus, these needs demonstrate the importance of the thorough characterization of shale gas reservoir (Table 4.1) as well as an understanding of how earth materials deform over various time scales and how it affect the current state of stress in the crust (Speight, 2014).

There is an additional issue that must be resolved if shale gas is to be a major contributor to US energy resources (or to the energy

Table 4.1 Variation in Shale Properties for Different Shale Gas Reservoirs (Sone, 2012)							
Sample Group	Estimated *In Situ* Stress (MPa)	Density (g/cc)	QFP (%)	Carbonate (%)	Clay (%)	Kerogen (%)	Porosity (%)
Barnett-1	Sv: 65 Pp: 30 σ_{eff}: 35	2.39−2.47	50−52	0−3	36−39	9−11	4−9
Barnett-2		2.63−2.67	31−53	37−60	3−7	2−3	1−2
Haynesville-1	Sv: 85 Pp: 60−70 σ_{eff}: 15−25	2.49−2.51	32−35	20−22	36−39	8−8	6−6
Haynesville-2		2.60−2.62	23−24	49−53	20−22	4−4	3−4
Eagle Ford-1	Sv: 90 Pp: 65 σ_{eff}: 25	2.43−2.46	22−29	46−54	12−21	9−11	0−3
Eagle Ford-2		2.46−2.54	11−18	63−78	6−14	4−5	3−5
Fort St. John	Sv: 25−30 Pp: 10−12 σ_{eff}: 13−20	2.57−2.60	54−60	3−5	32−39	4−5	5−6
QFP refers to quartz, feldspar, plagioclase, and pyrite component.							

resource plans of any country). The issue is the amenability of the gas to be included in current gas processing scenarios (Mokhatab et al., 2006; Speight, 2007). On the understanding that shale gas reservoirs will vary in properties such as origin, permeability, and porosity, differences in properties of the shale gas must be anticipated (Bustin et al., 2008) (see Chapters 1 and 2).

While shale gas resources represent a significant portion of current and future production, all shale gas is not constant in composition and gas processing requirements for shale gas can vary from area to area (Bullin and Krouskop, 2008; Weiland and Hatcher, 2012).

In addition, analysis of the gas composition of Devonian Shale wells indicates that the composition of produced gas shifts during the production history of the well (Schettler et al., 1989). Gas composition changes during production indicate that different components of natural gas produced have different decline curves. Thus, the total decline curve is the sum of the decline curves of the individual gas components. The classic mechanisms of viscous flow and ideal-gas void-volume storage, as such, do not explain gas fractionation in gas wells. In fact, the only way to explain observed fractionation with such classic mechanisms is to assume that total flow has several sources within the wellbore, each with different characteristic compositions and decline curves.

The flow rate of an individual component of a gas mixture can be obtained by multiplying the total flow rate by the mole fraction of the individual component. Even though the composition of gas from each source is assumed not to change, the composition of gas at the wellbore can change with time if the relative flows change with time (Schettler et al., 1989).

The second alternative is to assume that the gas composition changes observed at the wellbore reflect the changing composition of gas from at least some of the sources themselves. Candidates for such differences in gas composition include the presence of adsorption, solution, and/or diffusion. The occurrence of adsorption is associated with the presence of certain minerals in the reservoir, such as clay. Likewise, solution is associated with the presence of conventional petroleum or heavy oil or even gas condensate. The diffusion phenomenon is associated with diffusion through small pores such as those present in microporous reservoirs. Since these factors are commonly present, this second alternative may be involved in explaining the fractionation in many reservoirs (Schettler et al., 1989). Thus, as reservoir depletion occurs, the composition of the gas produced may approach the composition of the gas originally in the reservoir. Thus, composition shifts during production will be expected—but may not be expected, however, when the storage mechanism is associated solely with matrix porosity.

As a result of these various phenomena that can cause changes in gas composition, shale gas processors must be concerned about elevated ethane and nitrogen levels across a field. Other concerns are the increased requirements of urban gas processing. In addition, the rapid production growth in emerging shale areas can be difficult to handle.

Gas shale resources (see Chapter 2) represent a major contribution to the resource base of the United States. However, it is important to note that there is considerable variability in the quality of the resources, both within and between gas shale resources. Elevated levels of ethane, propane, carbon dioxide, or nitrogen in certain shale gases are of concern regarding their interchangeability with traditional natural gas supplies. This high level of variability in individual well productivity clearly has consequences with respect to the variability of individual well economic performance.

However, because shale is composed primarily of tiny grains of clay minerals and quartz, the mineral components of mud, the composition of the shale reservoir can vary—specifically the rock properties such as porosity, permeability, capillary entry pressure, pore volume compressibility, pore size distribution, and flow path (collectively known as *petrophysics*) can vary considerably (Sone, 2012). These materials were deposited as sediment in water, which was then buried, compacted by the weight of overlying sediment, and cemented together to form a rock (*lithification*). Clay minerals are a type of sheet silicate related to mica that usually occurs in the form of thin plates or flakes. As the sediment was deposited, the flakes of clay tended to stack together flat, one on top of another like a deck of cards, and as a result, lithified shale often has the property of splitting into paper-thin sheets. This is a convenient way to identify shale from other fine-grained rocks like limestone or siltstone.

This process, while appearing to be ordered in such a description, is in fact subject to geological disorder and thence subject to differing methods of entrapment of the organic material, different methods or rates of decomposition, different rates of formation of the gas, and hence different composition of the shale gas. The primary point of this section is that the geochemical and geological characteristics of each reservoir are relatively unique and must be carefully examined to determine resources. Furthermore, general rules are difficult to apply—innovation in unconventional drilling and completion techniques has added substantial reserves to otherwise uneconomic areas and has also been a key part of safe and efficient development (Cramer, 2008). This can affect the economics of shale gas production and gas cleaning (Mokhatab et al., 2006; Speight, 2007, 2014).

A major driver of shale economics is the amount of hydrocarbon liquid produced along with gas. Some areas contain wet gas with appreciable amounts of liquid, which can have a considerable effect on the breakeven economics, particularly if the price of oil is high compared to the price of gas.

The liquid content of a gas is often measured in terms of the *condensate ratio*, which is expressed in terms of barrels of liquid per million cubic feet of gas (bbls/MMcf). In some operations, for example in the case of a condensate ratio in excess of 50 bbls/MMcf, the liquid production alone can provide an adequate return on the investment, even if the gas cannot realize a fair market value.

4.2 GAS PROCESSING

Gas processing (gas treating, gas refining) usually involves the removal of (i) oil, (ii) water, (iii) elements such as sulfur, helium, and carbon dioxide, and (iv) natural gas liquids (see Chapter 6). In addition, it is often necessary to install scrubbers and heaters at or near the wellhead that serve primarily to remove sand and other large-particle impurities. The heaters ensure that the temperature of the natural gas does not drop too low and form a hydrate with the water vapor content of the gas stream (Mokhatab et al., 2006; Speight, 2007, 2014).

Many chemical processes are available for processing or refining natural gas. However, there are many variables in the process of refining sequence which dictate the choice of process or processes to be employed. In this choice, several factors must be considered: (i) the types and concentrations of contaminants in the gas, (ii) the degree of contaminant removal desired, (iii) the selectivity of acid gas removal required, (iv) the temperature, pressure, volume, and composition of the gas to be processed, (v) the carbon dioxide—hydrogen sulfide ratio in the gas, and (vi) the desirability of sulfur recovery due to process economics or environmental issues.

In addition to hydrogen sulfide and carbon dioxide, gas may contain other contaminants, such as mercaptans (also called *thiols*, RSH) and carbonyl sulfide (COS). The presence of these impurities may eliminate some of the sweetening processes since some processes remove large amounts of acid gas but not to a sufficiently low concentration. On the other hand, there exist processes that are not designed to remove (or are incapable of removing) large amounts of acid gases. However, these processes are also capable of removing the acid gas impurities to very low levels when the acid gases are present in low to medium concentrations in the gas.

Process selectivity indicates the preference with which the process removes one acid gas component relative to (or in preference to) another. For example, some processes remove both hydrogen sulfide and carbon dioxide; other processes are designed to remove hydrogen sulfide only. It is very important to consider the process selectivity for, say, hydrogen sulfide removal compared to carbon dioxide removal that ensures minimal concentrations of these components in the product; thus, there is a need for consideration of the carbon dioxide to hydrogen sulfide in the gas stream (Kohl and Riesenfeld, 1985; Maddox, 1982; Newman, 1985; Soud and Takeshita, 1994).

To include a description of all of the possible process for gas cleaning is beyond the scope of this book. Therefore, the focus of this chapter is a brief description of the processes that are an integral part within the concept of production of a pipeline-able product (methane) for sale to the consumer.

Gas processing equipment, whether in the field or at processing/ treatment plants, assures that these requirements can be met. While in most cases, processing facilities extract contaminants and higher-molecular weight hydrocarbons (natural gas liquids) from the gas stream, in a few cases, the higher molecular weight hydrocarbons may be blended into the gas stream to bring it within acceptable Btu levels.

Whatever the situation, there is the need to prepare the gas for transportation and use in domestic and commercial furnaces. Thus, natural gas processing begins at the wellhead and since the composition of the raw natural gas extracted from producing wells depends on the type, depth, and location of the underground deposit and the geology of the area, processing must offer several options (even though each option may be applied to a different degree) to accommodate the difference in composition of the extracted gas.

In those few cases where pipeline-quality natural gas is actually produced at the wellhead or field facility, the natural gas is moved directly to the pipeline system. In other instances, especially in the production of nonassociated natural gas, field or lease facilities referred to as *skid-mount plants* are installed nearby to dehydrate (remove water) and decontaminate (remove dirt and other extraneous materials) raw natural gas, so converting it into acceptable pipeline-quality gas for direct delivery to the pipeline system. The *skids* are often specifically customized to process the type of natural gas produced in the area and are a relatively inexpensive alternative to transporting the natural gas to distant large-scale plants for processing.

Gas processing (Mokhatab et al., 2006; Speight, 2007, 2014) consists of separating all of the various hydrocarbons, non-hydrocarbons (such as carbon dioxide and hydrogen sulfide), and fluids from the methane (Figure 4.1; Table 4.2). Major transportation pipelines usually impose restrictions on the makeup of the natural gas that is allowed into the pipeline. That means that before the natural gas can be transported it must be purified. While the ethane, propane, butanes, and pentanes must be removed from natural gas, this does not mean that they are all waste products.

Gas processing (gas refining) is necessary to ensure that the natural gas intended for use is clean burning and environmentally acceptable. Natural gas used by consumers is composed almost entirely of methane but natural gas that emerges from the reservoir at the wellhead is by no means *pure* (see Chapter 3). Although the processing of natural gas is in many respects less complicated than the processing and refining of crude oil, it is equally as necessary before its use by end users.

Raw natural gas comes from three types of wells: oil wells, gas wells, and condensate wells (see Chapters 2 and 4). *Associated gas*

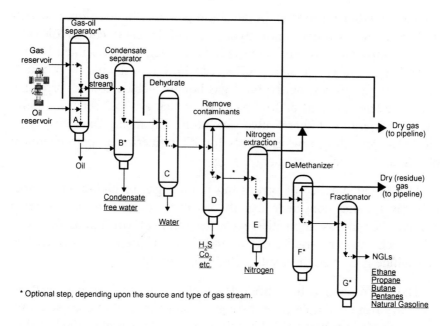

Figure 4.1 General schematic flow for gas processing. Energy Information Administration, Office of Oil and Gas, Natural Gas Division.

Table 4.2 Range of Composition of Natural Gas		
Methane	CH_4	70–90%
Ethane	C_2H_6	0–20%
Propane	C_3H_8	
Butane	C_4H_{10}	
Pentane and higher hydrocarbons	C_5H_{12}	0–10%
Carbon dioxide	CO_2	0–8%
Oxygen	O_2	0–0.2%
Nitrogen	N_2	0–5%
Hydrogen sulfide, carbonyl sulfide	H_2S, COS	0–5%
Rare gases: argon, helium, neon, xenon	A, He, Ne, Xe	Trace

(see Chapter 1), i.e., gas from petroleum wells, can exist separate from oil in the formation (free gas) or dissolved in the crude oil (dissolved gas). *Nonassociated gas*, i.e., gas from gas wells or condensate wells, is *free* natural gas along with a semiliquid hydrocarbon condensate. Whatever the source of the natural gas, once separated from crude oil (if present) it commonly exists in mixtures with other hydrocarbons, principally ethane, propane, butane, and pentane. In addition, raw natural gas contains water vapor, hydrogen sulfide (H_2S), carbon dioxide, helium, nitrogen, and other compounds. In fact, the associated hydrocarbons (*natural gas liquids*, NGLs) can be very valuable by-products of natural gas processing. Natural gas liquids include ethane, propane, butane, isobutane, and natural gasoline, which are sold separately and have a variety of different uses—including enhancing oil recovery in oil wells, providing raw materials for oil refineries or petrochemical plants, and as sources of energy.

The actual practice of processing natural gas to high-quality pipeline gas for the consumer usually involves four main processes to remove the various impurities: (i) water removal, (ii) liquids removal, (iii) enrichment, (iv) fractionation, and (v) the process by which hydrogen sulfide is converted to sulfur (the *Claus process*).

The processes that have been developed to accomplish gas purification vary from a simple once-through wash operation to complex multistep recycling systems (Mokhatab et al., 2006; Speight, 2007, 2014). In many cases, the process complexities arise because of the need for recovery of the materials used to remove the contaminants or even for the recovery of the contaminants in the original, or altered, form (Kohl and Riesenfeld, 1985; Mokhatab et al., 2006; Newman, 1985).

Initially, natural gas receives a degree of cleaning at the wellhead. The extent of the cleaning depends upon the specification that the gas must meet to enter the pipeline system. For example, natural gas from high-pressure wells is usually passed through field separators at the well to remove hydrocarbon condensate and water. Natural gasoline, butane, and propane are usually present in the gas, and gas processing plants are required for the recovery of these liquefiable constituents.

Throughout this chapter, two terms are used frequently: (i) absorption and (ii) adsorption.

Absorption is an approach in which the absorbed gas is ultimately distributed throughout the absorbent (liquid). The process depends only on physical solubility and may include chemical reactions in the liquid phase (*chemisorption*). Common absorbing media used are water, aqueous amine solutions, caustic sodium carbonate, and nonvolatile hydrocarbon oils, depending on the type of gas to be absorbed. Usually, the gas–liquid contactor designs which are employed are plate columns or packed beds.

Absorption is achieved by dissolution (a physical phenomenon) or by reaction (a chemical phenomenon). Chemical adsorption processes adsorb sulfur dioxide onto a carbon surface where it is oxidized (by oxygen in the flue gas) and absorbs moisture to give sulfuric acid impregnated into and on the adsorbent.

Adsorption differs from *absorption*, in that it is a physical–chemical phenomenon in which the gas is concentrated on the surface of a solid or liquid to remove impurities. Usually, carbon is the adsorbing medium (Mokhatab et al., 2006; Speight, 2007), which can be regenerated upon *desorption*. The quantity of material adsorbed is proportional to the surface area of the solid and, consequently, adsorbents are usually granular solids with a large surface area per unit mass. Subsequently, the captured gas can be desorbed with hot air or steam either for recovery or for thermal destruction.

The number of steps and the type of process used to produce pipeline-quality natural gas most often depend upon the source and makeup of the wellhead production stream. In some cases, several of the steps (Figure 4.1) may be integrated into one unit or operation, performed in a different order, or performed at alternative locations, or be not required at all (Mokhatab et al., 2006; Speight, 2007, 2014).

In many instances, pressure relief at the wellhead will cause a natural separation of gas from oil (using a conventional closed tank, where gravity separates the gas hydrocarbons from the higher-boiling crude oil). In some cases, however, a multistage gas-oil separation process is needed to separate the gas stream from the crude oil. The *gas-oil separators* used are commonly closed cylindrical shells, horizontally mounted with inlets at one end, an outlet at the top for removal of gas, and an outlet at the bottom for removal of oil. Separation is accomplished by alternately heating and cooling (by compression) the flow stream through multiple steps; some water and condensate, if present, will also be extracted as the process proceeds.

At some stage of the processing, the gas flow is directed to a unit that contains a series of filter tubes. As the velocity of the stream reduces in the unit, primary separation of remaining contaminants occurs due to gravity. Separation of smaller particles occurs as gas flows through the tubes, where they combine into larger particles which flow to the lower section of the unit. Furthermore, as the gas stream continues through the series of tubes, a centrifugal force is generated which further removes any remaining water and small solid particulate matter.

4.3 SHALE GAS

Shale gas refers to natural gas (mainly methane) found in fine-grained, organic-rich rocks (gas shale) (see Chapters 1 and 2). In addition, the descriptor word *shale* does not refer to a specific type of rock but, in addition to shale (mudstone), has also been used to describe rocks with more fine-grained particles (smaller than sand) than coarse-grained particles, such as (i) siltstone and (ii) fine-grained sandstone interlaminated with shale, as well as (iii) carbonate rocks. Thus, shale (including the additional rock types mentioned above) is a source rock that has not released all of the generated hydrocarbons. In fact, source rock that are *tight* or *inefficient* at expelling hydrocarbons may be the best prospects for shale gas potential. Thus, in gas shale, shale is a reservoir rock, source rock, and also a trap for natural gas. The natural gas found in these rocks is considered unconventional, similar to coalbed methane.

Shale gas is generated by any combination of (i) primary thermogenic degradation of organic matter, (ii) secondary thermogenic decomposition of petroleum, and (iii) biogenic degradation of organic matter. The respective gases generated by thermogenic and biogenic pathways may both exist in the same shale reservoir.

Gas is stored in shale in three different ways: (i) *adsorbed gas*, which is physically attached (adsorption) or chemically attached (chemisorption) to organic matter or to clay, (ii) *free gas* (also referred to as *nonassociated gas*; Speight, 2014), which occurs within the pore spaces in the rock or in spaces created by the rock cracking (fractures or microfractures), and (iii) solution gas (also referred to as *associated gas*; Speight, 2014), which exists in solution in liquids such as petroleum and heavy oil. The amount of adsorbed methane usually increases with an increase in organic matter or surface area of organic matter and/or clay.

Higher free-gas content in unconventional shale gas reservoirs generally results in higher initial rates of production because the free gas resides in fractures and pores and is easier to get out relative to adsorbed gas. The high, initial flow rates decline rapidly to a low, steady rate within about 1 year as adsorbed gas is slowly released from the shale.

Compared to most conventional reservoirs (typically a *sandstone reservoir*), the gas shale reservoir has an extremely low permeability. In fact, the effective bulk permeability in gas shale is typically much less than 0.1 mD, although exceptions exist where the rock is naturally fractured: for example, the well-fractured Antrim Shale in the Michigan Basin in the United States. In most cases, it is normal for the well to need artificial stimulation, such as fracturing, to increase permeability to the well. This helps the well to produce gas in economical quantities. The role of natural microfractures in reservoir production or in assisting artificial fracturing is not well understood.

4.4 SHALE GAS PROPERTIES AND PROCESSING

The shale formations in the United States that presently produce gas commercially exhibit an unexpectedly wide variation in the values of five key parameters: (i) thermal maturity, expressed as vitrinite reflectance, (ii) sorbed gas fraction, (iii) reservoir thickness, (iv) total organic carbon (TOC) content, and (v) volume of gas in place (see Chapter 2). In addition, the degree of natural fracture development in an otherwise low-matrix-permeability shale reservoir is a controlling factor in gas producibility and, possibly, in gas properties.

A wide range of reservoir properties (see Chapter 2) control both the rate and the volume of shale gas production from the gas shale formations, notably: (i) thermal maturity, (ii) gas in place, TOC content, (iii) reservoir thickness, and (iv) proportion of sorbed gas. Natural fracture networks are required to augment the extremely low shale matrix permeability. Therefore, geology and geochemistry must be considered together when evaluating a given shale system both before and after drilling as well as gas processing options.

In addition, it is very likely that not only the amount and distribution of gas within the shale but also the composition of the shale gas is determined by, among other things: (i) the initial reservoir pressure, (ii) the petrophysical properties of the rock, and (iii) the adsorption characteristics of the rock—thus, during production there are three main processes that can be operative.

Initial gas production is dominated by depletion of gas from the fracture network. This form of production declines rapidly due to limited storage capacity. Once the initial decline rate stabilizes, the depletion of gas stored in the matrix becomes the primary process involved in production. The amount of gas held in the matrix is dependent on the particular properties of the shale reservoir, which can be hard to estimate. Secondary to this depletion process is desorption whereby adsorbed gas is released from the rock as pressure in the reservoir declines. The rate of gas production via the desorption process depends on there being a significant drop in reservoir pressure. At the same time, the composition of the gas can, and undoubtedly does, change due to the action of any one of these parameters or due to the interaction of any two of the above parameters or the interaction of all three of the above parameters.

Furthermore, in the Western Canadian Sedimentary Basin (WCSB), Devonian–Mississippian (D–M) and Jurassic Shale formations have complex, heterogeneous pore volume distributions as identified by low-pressure carbon dioxide and nitrogen sorption (Ross and Bustin, 2009). In fact, high-pressure methane isotherms on dried and moisture equilibrated shale samples show a general increase of gas sorption with TOC content. Methane sorption in D–M formations increases with increasing TOC content and micropore volume, indicating that microporosity associated with the organic fraction is a primary control not only on methane sorption but also on shale gas composition.

The sorption capacities for Jurassic Shale formations may be unrelated to micropore volume and the large sorbed gas capacities of organic-rich Jurassic Shale formations, independent of surface area, imply a portion of the methane is stored by solution in matrix bituminite (Ross and Bustin, 2009). Solute methane does not appear to be an important contributor to gas storage in D–M shale formations (Ross and Bustin, 2009). In fact it is likely that structural transformation of D–M organic matter has occurred during thermal diagenesis creating and/or opening up microporosity onto which gas can sorb thereby also influencing the composition of the shale gas.

Furthermore, inorganic constituents also influence modal pore size, total porosity, and sorption characteristics of shale formation, thereby adding a further parameter to those parameters (above) that influence changes in the shale gas composition. Clays are known to provide excellent adsorption surfaces for petroleum constituents (Speight, 2014) and are also capable of sorbing gas into the internal structure, the amount of which is dependent on clay type.

The uncertainties of reservoir properties and fracture parameters (see Chapters 2 and 3) have a significant effect on shale gas properties and production, making the process of optimization of hydraulic fracturing treatment design for economic gas production much more complex. It is extremely important to identify reasonable ranges for these uncertainty parameters and evaluate their effects on well performance, because the detailed reservoir properties for each wellbore are difficult to determine.

Gas production from unconventional shale gas reservoirs has become more common in the past decade, and there are increasing demands to understand the petrophysical and mechanical properties of these rocks. Characterizing these organic-rich shale formations can be challenging as these rocks vary quite significantly (Passey et al., 2010). For instance, formations in the Barnett Shale are known to be more silica-rich, whereas Eagle Ford Shale rocks are generally carbonate-rich containing relatively smaller amounts of silica and clay. Formations in these shale gas reservoirs also exhibit a wide range of composition within a single reservoir. There are also indications that it is not only the amount of clay or organics, but also the maturity of the shale formations that control the anisotropy of these organic-rich shale formations (Vanorio et al., 2008).

Not surprisingly, gas-bearing shale formations are complex reservoirs (with a porosity on the order of 4 to 6 porosity units and a total organic carbon content of $\geq 4\%$ w/w), which represent significant variety in reservoir characteristics (i.e., mineralogy, porosity, permeability, gas content, and pressure). In addition, the shale porosity changes at very different rates in different regions and formations. Moreover, the gas in these shale formations occurs both as a free phase within pores and fractures and as gas sorbed onto organic matter. Not surprisingly, and in accordance with the varying shale reservoir properties, it must be anticipated that there will be differences in shale gas composition and properties.

Thus, although shale gas represents a large, new source of natural gas and natural gas liquids (NGLs), shale gas is certainly not the same everywhere. Produced shale gas observed to date has shown a broad variation in compositional makeup, with some having wider component ranges, a wider span of minimum and maximum heating values, and higher levels of water vapor and other substances than pipeline tariffs or purchase contracts may typically allow. Indeed, because of these variations in gas composition, each shale gas formation can have unique processing requirements for the produced shale gas to be marketable. Ethane can be removed by cryogenic extraction while carbon dioxide can be removed through a scrubbing process. However, it is not always necessary (or practical) to process shale gas to make its composition identical to that of *conventional* transmission-quality gases. Instead, the gas should be interchangeable with other sources of natural gas now provided to end users. The interchangeability of shale gas with conventional gases is crucial to its acceptability and eventual widespread use in the United States.

Although not highly sour in the usual sense of having high hydrogen sulfide content, and with considerable variation from resource to resource and even from well to well within the same resource (due to extremely low permeability of the shale even after fracturing) (see Chapters 1 and 3), shale gas often contains varying amounts of hydrogen sulfide with wide variability in the carbon dioxide content. The gas is not ready for pipelining immediately after it has exited the shale formation.

The challenge in treating such gases is the low (or differing) hydrogen sulfide/carbon dioxide ratio and the need to meet pipeline specifications. In a traditional gas processing plant, the olamine of choice for hydrogen sulfide removal is *N*-methyl diethanolamine (MDEA) (Mokhatab et al., 2006; Speight, 2007) but whether or not this olamine will suffice to remove the hydrogen sulfide without removal of excessive amounts of carbon dioxide is another issue.

Gas treatment may begin at the wellhead—condensates and free water usually are separated at the wellhead using mechanical separators. Gas, condensate, and water are separated in the field separator. Extracted condensate and free water are directed to separate storage tanks and the gas flows to a gathering system. After the free water has been removed, the gas is still saturated with water vapor, and depending on the temperature and pressure of the gas stream, may need to be dehydrated or treated with methanol to prevent hydrates forming as the temperature drops. But this may not be always the case in actual practice.

Thus, there exists the real need to evaluate gas processing operations and the ability of a processing plant to treat a variety of shale gases to pipeline specifications. Solvent selection, strength, temperature, and circulation rate, as well as the type and quantity of internals used in the contactor, are some of the process parameters and design variables that *must* be considered (Weiland and Hatcher, 2012).

4.4.1 Antrim Shale Formation

The Antrim Shale is a shallow shale gas resource in Michigan. The Antrim Shale is unique because the gas is predominantly biogenic: methane is created as a by-product of bacterial consumption of organic material in the shale. The carbon dioxide level in these samples varies from 0 to 9%v/v (Table 4.3). Carbon dioxide is a naturally occurring by-product of shale gas produced by desorption and, as a result, the carbon dioxide levels in shale gas increase during the productive life of a well.

Individual well production varies from 50,000 to 60,000 ft^3/day. Significant associated water is produced, which requires central production facilities for dehydration, compression and disposal (Bullin and Krouskop, 2008).

The Antrim Shale gas, for example, has high nitrogen concentration, as does at least one well tested in the Barnett Shale formation. New Albany Shale gas show high carbon dioxide concentrations while several wells in the Marcellus Gas Shale have tested up to 16% v/v. Economically treating and processing these gases requires all the same techniques as are used for conventional gas—plus the ability to handle a great deal of variability in the same field. These differences in quality of shale gas introduce a note of caution into gas processing because of the variability of the composition and properties of the gas (Bullin and Krouskop, 2008; Weiland and Hatcher, 2012).

4.4.2 Barnett Shale Formation

The Barnett Shale formation is the more familiar of shale gas resources and grandfather of shale gas plays. The Barnett Shale formation lies around the Dallas–Fort Worth area of Texas and produces gas at depths of 6500–9500 ft.

The initial discovery region was in a core area on the eastern side of the play. As drilling has moved westward, the form of the hydrocarbons in the Barnett Shale has varied from dry gas prone in the east to oil prone in the west (Table 4.3). As a result of such variations in composition, blending may be the most appropriate methods for equalizing the variations. With the richness of the gas, the Barnett Plants remove substantial amounts of natural gas liquids each day.

The Barnett Shale formation is the most well known of shale gas formations (see Chapter 2). Much of the technology used in drilling and production of shale gas has been developed for this resource (Bullin and Krouskop, 2008; Weiland and Hatcher, 2012). The Barnett Shale formation and produces at depths of 6500−9500 ft with a production rate on the order of 0.5 to 4 million cubic feet per day (ft^3/day) with estimates of 300−550 ft^3 of gas per ton of shale.

The initial discovery region was in a core area on the eastern side of the resource, and as the drilling and gas recovery have moved westward, the composition of the shale gas has changed from *dry gas* production in the east to *wet gas* and *oil* production in the west.

The Barnett Shale resource play of North Texas, for example, contains several hundred parts per million (ppm v/v) of hydrogen sulfide and much higher amounts of carbon dioxide (in the percent v/v range). In other shale resources, such as Haynesville and the Eagleville field (Eagle Ford resource), a hydrogen sulfide content is known to be present. In other gas shale resources, such as the Antrim resource and the New Albany resource, underlying Devonian formations may communicate with and contaminate the shale formations. In addition, some of the shale gas resources have low carbon dioxide content but have a sufficiently high content of hydrogen sulfide to require treating. Thus, even after removing the natural gas liquids, there are reasons for the shale gas to require further treatment to remove hydrogen sulfide and carbon dioxide to meet pipeline specifications.

4.4.3 Fayetteville Shale Formation

The Fayetteville Shale is an unconventional gas reservoir located on the Arkansas side of the Arkoma Basin and ranges in thickness from 50 to 550 ft at a depth of 1500−6500 ft. The gas (Table 4.3) primarily requires only dehydration to meet pipeline specifications.

The shale formation is estimated to hold between 58 and 65 billion cubic feet (58 to 65 × 10^6 ft^3) per square, and initial production rates were on the order of 0.2−0.6 million cubic feet per day for vertical

Well	C_1	C_2	C_3	CO_2	N_2
Table 4.3 Variation of Shale Gas Composition with Well Placement (Bullin and Krouskop, 2008; Hill et al., 2007; Martini et al., 2003)					
Antrim Shale Gas					
1	27.5	3.5	1.0		65.0
2	57.3	4.9	1.9	0.3	5.9
3	77.5	4.0	0.9	3.3	14.3
4	85.6	4.3	0.4	9.0	0.7
Barnett Shale Gas					
1	80.3	8.1	2.3	1.4	7.9
2	81.2	11.8	5.2	0.3	1.5
3	91.8	4.4	0.4	2.3	1.1
4	93.7	2.6	0.0	2.7	1.0
Fayetteville Shale Gas					
1	97.3	1.0	0	1.0	0.7
Haynesville Shale Gas					
1	95.0	0.1	0	4.8	0.1
Marcellus Shale Gas					
1	79.4	16.1	4.0	0.1	0.4
2	82.1	14.0	3.5	0.1	0.3
3	83.8	12.0	3.0	0.9	0.3
4	95.5	3.0	1.0	0.3	0.2
New Albany Shale Gas					
1	87.7	1.7	2.5	8.1	
2	88.0	0.8	0.8	10.4	
3	91.0	1.0	0.6	7.4	
4	92.8	1.0	0.6	5.6	

The compositions have been normalized to the reported compounds.

wells and 1 to 3.5 million cubic feet per day for horizontal wells. In 2003, the production era exceeded 500 million cubic feet per day (Bullin and Krouskop, 2008).

4.4.4 Haynesville Shale Formation

The Haynesville Shale gas resource is the newest shale area to be developed. It lies in northern Louisiana and East Texas. The formation is deep (>10,000 ft), with high bottomhole temperature (175°C, 350°F), and high pressure (3000–4000 psi).

The wells showed initial production rates up to twenty or more million cubic feet of gas per day with estimates of $100-330$ ft^3 of gas per ton of shale (Bullin and Krouskop, 2008).

The gas requires treating for carbon dioxide removal (Table 4.3). Operators in this field are using amine treating to remove the carbon dioxide with a scavenger treatment on the tail gas to remove hydrogen sulfide.

4.4.5 Marcellus Shale Formation

The Marcellus Shale lies in western Pennsylvania, Ohio, and West Virginia. The formation is shallow at depths of $2000-8000$ ft and is $300-1000$ ft thick. The gas composition varies across the field, much as it does in the Barnett—the gas becomes richer from east to west (Table 4.3). Initial production rates have been reported to be on the order of $0.5-4$ million cubic feet per day (ft^3/day), with estimates of $60-100$ cubic feet of gas per ton of shale (Bullin and Krouskop, 2008).

The Marcellus Shale gas has relatively little carbon dioxide and nitrogen. In addition, the gas is *dry* and does not require removal of natural gas liquids for pipeline transportation. Early indications are that the Marcellus gas has sufficient liquids to require processing.

4.4.6 New Albany Shale Formation

The New Albany Shale is black shale in Southern Illinois extending through Indiana and Kentucky. The formation is $500-4900$ ft deep and $100-400$ ft thick. The gas composition (Table 4.3) is variable and low flow rates of wells in the New Albany Shale require that the production from many wells must be combined to warrant processing the gas.

Vertical wells typically produce $25,000-75,000$ ft^3/day, whereas horizontal wells can have initial production rates up to 2 million cubic feet per day (Bullin and Krouskop, 2008).

REFERENCES

Ahmadov, R., Vanorio, T., Mavko, G., 2009. Confocal laser scanning and atomic-force microscopy in estimation of elastic properties of the organic-rich Bazhenov formation. Leading Edge 28, 18–23.

Bullin, K., Krouskop, P., 2008. Composition variety complicates processing plans for US shale gas. Proceedings of the Annual Forum, Gas Processors Association—Houston Chapter, October 7, Houston, TX.

Bustin, R.M., Bustin, A.M.M., Cui, X., Ross, D.J.K., Pathi, V.S.M., 2008. Impact of shale properties on pore structure and storage characteristics. Paper No. SPE 119892. Proceedings of the SPE Shale Gas Production Conference, November 16–18, Fort Worth, TX.

Cramer, D.D., 2008. Stimulating unconventional reservoirs: lessons learned, successful practices, areas for improvement. SPE Paper No. 114172. Proceedings of the Unconventional Gas Conference, February 10–12, 2008, Keystone, CO.

Hill, R.J., Jarvie, D.M., Zumberge, J., Henry, M., Pollastro, R.M., 2007. Oil and gas geochemistry and petroleum systems of the Fort Worth Basin. Am. Assoc. Pet. Geol. Bull. 91 (4), 445–473.

Kohl, A.L., Riesenfeld, F.C., 1985. Gas Purification, fourth ed. Gulf Publishing Company, Houston, TX.

Kundert, D., Mullen, M., 2009. Proper evaluation of Shale Gas reservoirs leads to a more effective hydraulic-fracture stimulation. Paper No. SPE 123586. Proceedings of the SPE Rocky Mountain Petroleum Technology Conference, April 14–16, Denver, CO.

Maddox, R.N., 1982. Gas and Liquid Sweetening, Gas Conditioning and Processing, vol. 4. Campbell Publishing Co., Norman, OK.

Martini, A.M., Walter, L.M., Ku, T.C.W., Budai, J.M., McIntosh, J.C., Schoell, M., 2003. Microbial production and modification of gases in sedimentary basins: a geochemical case study from a Devonian Shale Gas play, Michigan Basin. Am. Assoc. Pet. Geol. Bull. 87 (8), 1355–1375.

Martini, A.M., Walter, L.M., McIntosh, J.C., 2008. Identification of microbial and thermogenic gas components from upper Devonian Black Shale cores, Illinois and Michigan Basins. Am. Assoc. Pet. Geol. Bull. 92 (3), 327–339.

Mokhatab, S., Poe, W.A., Speight, J.G., 2006. Handbook of Natural Gas Transmission and Processing. Elsevier, Amsterdam, the Netherlands.

Newman, S.A., 1985. Acid and Sour Gas Treating Processes. Gulf Publishing, Houston, TX.

Passey, Q.R., Bohacs, K.M., Esch, W.L., Klimentidis, R., Sinha, S., 2010. From oil-prone source rocks to gas-producing shale reservoir—geologic and petrophysical characterization of unconventional shale-gas reservoir. SPE Paper No. 131350. Proceedings of the CPS/SPE International Oil & Gas Conference and Exhibition, June 8–10, Beijing, China.

Ross, D.J.K., Bustin, R.M., 2009. The importance of shale composition and pore structure upon gas storage potential of shale gas reservoirs. Mar. Pet. Geol. 26 (6), 916–927.

Schettler Jr., P.D., Parmely, C.R., Juniata, C., 1989. Gas composition shifts in Devonian Shales. SPE Reservoir Eng. 4 (3), 283–287.

Sone, H., 2012. Mechanical Properties of Shale Gas Reservoir Rocks and Its Relation to the In-Situ Stress Variation Observed in Shale Gas Reservoirs. A Dissertation Submitted to the Department of Geophysics and the Committee on Graduate Studies of Stanford University in Partial Fulfillment of the Requirements for the Degree of Doctor of Philosophy. SRB Volume 128, Stanford University, Stanford, CA.

Soud, H., Takeshita, M., 1994. FGD Handbook. International Energy Agency Coal Research, London, England (No. IEACR/65).

Speight, J.G., 2007. Natural Gas: A Basic Handbook. GPC Books, Gulf Publishing Company, Houston, TX.

Speight, J.G., 2014. The Chemistry and Technology of Petroleum, fifth ed. CRC Press, Taylor & Francis Group, Boca Raton, FL.

Vanorio, T., Mukerji, T., Mavko, G., 2008. Emerging methodologies to characterize the rock physics properties of organic-rich shales. Leading Edge 27, 780–787.

Weiland, R.H., Hatcher, N.A., 2012. Overcome challenges in treating shale gas. Hydrocarbon Process. 91 (1), 45–48.

CHAPTER 5

Environmental Issues

5.1 INTRODUCTION

Natural gas production from hydrocarbon-rich shale formations (shale gas) is one of the most rapidly expanding trends in onshore domestic oil and gas exploration and production. In some areas, this has included bringing drilling and production to regions of the country that have seen little or no activity in the past. New oil and gas developments bring change to the environmental and socioeconomic landscape, particularly in those areas where gas development is a new activity. With these changes have come questions about the nature of shale gas development, the potential environmental impacts, and the ability of the current regulatory structure to deal with this development. Regulators, policy makers, and the public need an objective source of information on which to base answers to these questions and decisions about how to manage the challenges that may accompany shale gas development.

Environmental impacts associated with shale gas development occur at the global and local levels. These include impacts to climate change (Schrag, 2012; Shine, 2009), local air quality, water availability, water quality, seismicity, and local communities (Clark et al., 2012).

The rapid expansion in shale gas production has given rise to concerns around the impact of operations in areas such as water, road, air quality, seismic, and greenhouse gas (GHG) emissions (Howarth et al., 2011a,b; O'Sullivan and Paltsev, 2012; Stephenson et al., 2011). The process of hydraulic fracturing (fracking) in a shale gas well requires significant volumes of water and causes additional GHG emissions compared to conventional gas wells (Spellman, 2013). There is already significant resistance to shale gas development due to these water and emission concerns in many parts of the United States and Western Europe, with some countries imposing a nationwide moratorium on shale gas production through fracking. The regulation of shale gas is an evolving issue as the industry has developed so rapidly that it has often outpaced the availability of information for regulators to develop specific guidance (Clark et al., 2012).

There are environmental concerns regarding the specialized techniques used to exploit shale gas (Arthur et al., 2008, 2009; GAO, 2012). There is potential for a heavy draw on freshwater resources because of the large quantities required for hydraulic fracturing fluid. The land-use footprint of shale gas development is not expected to be much more than the footprint of conventional operations, despite higher well densities, because advances in horizontal drilling technology allow for up to 10 or more wells to be drilled and produced from the same well site. Finally, there is potential for a high carbon footprint through emissions of carbon dioxide (CO_2), a natural impurity in some shale gas.

In fact, environmental concerns about shale gas exploitation have received significant attention in the media. The issues raised are freshwater usage in competition with other uses such as farming, improper disposal of produced water, and contamination of freshwater aquifers.

Although shale gas wells use up to 6 million gallons (6×10^6 gallons) of water per well, the water volume used per unit of energy produced is small compared to a number of alternatives. Although this usage is relatively low compared to alternatives, any usage of water may appear to be in competition with other uses, especially in drought years. To address this situation, salt water might be used in place of freshwater. Recent advances in fracturing permit this with small modifications to the needed chemicals.

Improper discharge of produced water is an issue. It is best addressed by simply recycling. However, since the produced water has salinity from 6000 to 300,000 ppm, this can be costly. The ability to tolerate salinity mentioned above can be a huge cost saving. Technology to clean up the other constituents exists and can be expected to be affordable.

There are two potential ways in which shale gas operations could contaminate aquifers. One is through leakage of the chemicals used in fracturing. These then would be liquid contaminants. The second is the infiltration of aquifers by produced methane. It is a gaseous contaminant, albeit it gets dissolved in the water.

If methane is present, a portion may be released as a gas. The distinction between potential liquid and gaseous contamination is important because the hazards are different, as are the remedies and safeguards.

Also, because well water could not naturally have the liquid contaminants, their presence is evidence of a man-made source.

Therefore, simple testing of wells proximal to drilling operations is sufficient, with the only possible complication being the influence from some source other than drilling, such as agricultural runoff. This is easily resolved because of the specificity in the chemicals used for fracturing.

Methane leakage can happen because of possible combination of not locating cement in the right places and of a poor cement job. Many wells will have intervals above the producing zone that are charged with gas, usually small quantities in coal bodies and the like. If these are not sealed off with cement, some gas will intrude into the wellbore. This will still be contained unless the cement up near the freshwater aquifers has poor integrity. In that case, the gas will leak. Wells constructed to specification will not leak.

5.2 ENVIRONMENTAL REGULATIONS

A series of federal laws govern most environmental aspects of shale gas development. For example, the *Clean Water Act* regulates surface discharges of water associated with shale gas drilling and production, as well as storm water runoff from production sites. The *Safe Drinking Water Act* (SDWA) directs the underground injection of fluids from shale gas activities. The *Clean Air Act* limits air emissions from engines, gas processing equipment, and other sources associated with drilling and production. The *National Environmental Policy Act* (NEPA) requires that exploration and production on federal lands be thoroughly analyzed for environmental impacts.

However, federal agencies do not have the resources to administer all of these environmental programs for all the oil and gas sites around the country. In addition, federal regulation may not always be the most effective way of assuring the desired level of environmental protection. Therefore, most of these federal laws have provisions for granting primacy to the states, which have usually developed their own sets of regulations. By statute, states may adopt these standards of their own, but they must be at least as protective as the federal principles they replace—they may actually be more protective in order to address local conditions.

Once these programs are approved by the relevant federal agency (usually the US Environmental Protection Agency), the state then has primacy jurisdiction. State regulation of the environmental practices related to shale gas development can more easily address the regional and state-specific character of the activities, compared to one-size-fits-all management at the federal level. Some of these factors include geology, hydrology, climate, topography, industry characteristics, development history, state legal structures, population density, and local economics.

Thus, the regulation of shale gas drilling and production is a cradle-to-grave approach, and states have many tools at their disposal to assure that shale gas operations do not adversely impact the environment. They have broad powers to regulate, permit, and enforce all activities—from drilling and fracturing of the well, to production operations, to managing and disposing of wastes, to abandoning and plugging the production well(s).

Different states take different approaches to this regulation and enforcement, but their laws generally give the state oil and gas director or the agency the discretion to require whatever is necessary to protect the human health and the environment. In addition, most have a general prohibition against pollution from oil and gas drilling and production. A majority of the state requirements are written into rules or regulations; however, some are added to permits on a case-by-case basis as a result of environmental review, on-the-ground inspections, public comments, or commission hearings.

Finally, the organization of regulatory agencies within the different oil-and-gas-producing states varies considerably. Some states have several agencies that may oversee some facet of oil and gas operations, particularly environmental requirements. In different states, these agencies may be located in sundry departments or divisions within their respective governments. These various approaches have developed over time within each state, and each state tries to create a structure that best serves its citizenry and all of the industries that it must oversee. The one constant is that each oil-and-gas-producing state has one agency with primary responsibility for permitting wells and overseeing general operations. While these agencies may work with other agencies in the regulatory process, they serve as a central organizing body and a useful source of information about the various agencies that may have jurisdiction over oil and gas activities (Arthur et al., 2008).

5.2.1 General Aspects

Although hydraulic fracturing is not directly regulated by federal standards, a number of federal laws still direct oil and gas development, including shale gas (Spellman, 2013; US EPA, 2012). These regulations affect water management and disposal, as well as air quality and activities on federal land (Gaudlip et al., 2008; Veil, 2010). The *Clean Water Act* is focused on surface waters and regulates disposal of wastewater and also includes authorizing the *National Pollutant Discharge Elimination System* (NPDES) permit program, as well as requiring tracking of any toxic chemicals used in fracturing fluids. The *Hazardous Materials Transport Act* and the *Oil Pollution Act* both regulate ground pollution risks relating to spills of materials or hydrocarbons into the water table.

Regulation of oil and gas production has traditionally occurred primarily at the state level, which is currently also the case for shale gas, with most shale gas-producing states issuing more rigorous standards that take primacy over federal regulations, as well as additional regulations that control areas not covered at the federal level, such as hydraulic fracturing. Within states, regulation is carried out by a range of agencies.

Energy or natural resource-focused departments generally set requirements for site permits, drilling, completion, and extraction, while environmental or water departments regulate water emissions and waste management. The specific regulations vary considerably among states, such as different depths for well casing, levels of disclosure on drilling and fracturing fluids, or requirements for water storage. The majority of states in shale gas-producing regions now have varying hydraulic fracturing regulations on their books, specifically for disclosure of fracking fluids, proper casing of wells to prevent aquifer contamination, and management of wastewater from flowback and produced water. Disposal of wastewater by underground injection has emerged as a point of concern for state regulators due to large interstate flows of wastewater to states with suitable geology and reports of seismic activity near some well sites.

5.2.2 New Regulations

Hydraulic fracturing techniques have grown to be the carefully engineered processes employed to generate a more extensive network of fractures and thereby produce a larger portion of the natural gas in-place. This innovation has transformed shale gas into a bona fide economic resource play and has led to the drilling of many more shale gas wells and to increased attention on potential environmental effects.

Briefly and historically, hydraulic fracturing of gas wells began in 1949; however, it remained largely unregulated until significant unconventional gas production began at the beginning of the twenty-first century with the commercial development of coalbed methane. As production grew, reports of drinking water contamination raised concerns, leading the US Environmental Protection Agency to commission a study into the risks of hydraulic fracturing to drinking water. In 2004, this study found that hydraulic fracturing of coalbed methane posed minimal threat to underground sources of drinking water, which was a significant finding in support of the industry. In 2005, the federal *Energy Policy Act* granted hydraulic fracturing a specific exemption from the SDWA, which regulates all underground injection.

Since the *Energy Policy Act* passed in 2005, shale gas production in the United States has grown significantly, from less than one trillion cubic feet $(1.0 \times 10^{12} \text{ ft}^3)$ in 2005 to more than three trillion cubic feet $(3.0 \times 10^{12} \text{ ft}^3)$ in 2009. Such rapid growth, along with continued reports of environmental effects, has led to renewed calls for the federal government to provide increased regulation or guidance. This pressure led to the introduction to Congress in 2009 of the *Fracturing Responsibility and Awareness of Chemicals Act* (the FRAC Act) to define hydraulic fracturing as a federally regulated activity under the SDWA. The proposed act requires the energy industry to disclose the chemical additives used in the hydraulic fracturing fluid. The act did not receive any action, was reintroduced in 2011, and appears to have been a nonissue since that time.

In the absence of new federal regulation, states have continued to use existing oil and gas and environmental regulations to manage shale gas development, as well as introducing individual state regulations for hydraulic fracturing. In fact, the current regulation is comprised of an overlapping collection of federal, state, and local regulations and permitting systems, implemented by oil and gas, natural resources, and environmental agencies.

These regulations cover different aspects of the development and production of a shale gas well, with the intention that they combine to manage any potential impact on the surrounding environment and water supplies. These combinations of regulations have long served to regulate oil and gas development in numerous states. However, the new process of hydraulic fracturing is something that has not previously been managed by these regulations.

Therefore, the related intensity in terms of water, emissions, and site activity means existing regulations are being reassessed for their suitability for this new production method.

In the meantime, many states (including Wyoming, Arkansas, and Texas) have already implemented regulations requiring disclosure of the materials used in fracking fluids, and the US Department of the Interior has indicated an interest in requiring similar disclosure for sites on federal lands.

The Bureau of Land Management (BLM) of the Department of the Interior proposed draft rules for oil and gas production on public lands require disclosure of the chemical components used in hydraulic fracturing fluids, among other groundwater protections. The proposed rule requires the operator to submit an operation plan prior to hydraulic fracturing that would allow the BLM to evaluate groundwater protection designs based on the local geology, review anticipated surface disturbance, and approve proposed management and disposal of recovered fluids. In addition, operators would provide to the BLM the information necessary to confirm wellbore integrity before, during, and at the conclusion of the stimulation operation. Before hydraulic fracturing begins, operators would have to self-certify that the fluids comply with all applicable federal, state, and local laws and rules and regulations. After the conclusion of hydraulic fracturing, a follow-up report would summarize what actually occurred during fracturing activities, including the specific chemical makeup of the hydraulic fracturing fluid.

In addition to EPA's regulatory authority under the SDWA, EPA is exploring the possibility of developing rules under the *Toxic*

Substances Control Act (TSCA) to regulate the reporting of hydraulic fracturing fluid information. The EPA also has authority under the *Clean Air Act* to regulate hazardous air emissions from hydraulic fracturing operations.

On April 17, 2012, EPA released new source performance standards and national emissions standards for hazardous air pollutants (HAPs) in the oil and natural gas sector. The final rules include the first federal air standards for hydraulically fractured gas wells, along with requirements for other sources of pollution in the oil and gas industry that currently are not regulated at the federal level. These standards require either flaring or green completions on all feasible natural gas wells developed prior to January 1, 2015, with only green completions allowed for wells developed on and after that date. These rules are expected to reduce emissions of volatile organic compounds (VOCs) from applicable hydraulically fractured wells by approximately 95%, while reducing the emissions of VOCs, HAPs, and methane by approximately 10%.

5.3 ENVIRONMENTAL IMPACT

It is important to recognize the inherent risks of the oil and gas business and the damage that can be caused by just one poor operation; the industry must continually strive to mitigate risk and address public concerns. Particular attention should be paid to those areas of the country which are not accustomed to oil and gas development and where all relevant infrastructure, both physical and regulatory, may not yet be in place (Arthur et al., 2008).

The fracturing process (see Chapter 3) entails the pumping of fracture fluids, primarily water with sand proppant and chemical additives, at sufficiently high pressure to overcome the compressive stresses within the shale formation for the duration of the fracturing procedure. Each stage is typically of the order of a few hours. The process increases formation pressure above the critical fracture pressure, creating narrow fractures in the shale formation. The sand proppant is then pumped into these fractures to maintain a permeable pathway for fluid flow after the fracture fluid is withdrawn and the operation is completed.

While there are several environment-related issues that must receive attention, the major focus has been on the fracturing process, which poses risk to the shallow groundwater zones that may exist in close proximity to the gas-bearing formation.

As described previously, multiple layers of cement and casing protect the freshwater zones as the fracture fluid is pumped from the surface down into the shale formation. This protection is tested at high pressures before the fracturing fluids are pumped downhole. Once the fracturing process is underway, the large vertical separation between the shale sections being fractured and the shallow zones prevents the growth of fractures from the shale formation into shallow groundwater zones. It should be noted here that only shallow zones contain potable water; as depths increase, the salinity of the groundwater increases to the point that it has no practical utility.

The primary risks are (i) contamination of groundwater aquifers with drilling fluids or natural gas while drilling and setting casing through the shallow zones, (ii) on-site surface spills of drilling fluids, fracture fluids, and wastewater from fracture flowbacks, (iii) contamination as a result of inappropriate off-site wastewater disposal, (iv) excessive water withdrawals for use in high-volume fracturing, and (v) excessive road traffic and impact on air quality.

5.3.1 Air Pollution

Shale gas production activities can produce significant amounts of air pollution that could impact local air quality in areas of concentrated development. In addition to GHG emissions, fugitive emissions of natural gas can release VOCs and HAPs, such as benzene. Nitrogen oxides (NOx) are another pollutant of concern, as drilling, hydraulic fracturing, and compression equipment—typically powered by large internal combustion engines—produce these emissions.

Several state emission inventories have shown that oil and natural gas operations are significant sources of local air pollution. However, uncertainty about the impacts of these emissions exists, as air quality is highly dependent on local conditions. For example, in some areas, emissions of volatile organic compounds will not be the primary driver of ozone formation; therefore, detailed modeling is required to understand the impact of emissions on local air quality. In addition, while elevated levels of benzene emissions have been found near production sites, concentrations have been below health-based screening levels, and with little data on how the emissions of hazardous air pollutants impact human health, further examination is needed.

GHG emissions and other air emissions from shale well sites are also a key environmental concern. GHG emissions are generated in shale gas operations from exploration through processing to transmission and distribution.

The US Environmental Protection Agency has finalized GHG emissions reporting regulations from many of these emissions sources under the Mandatory Reporting of Greenhouse Gases Rule. Additional air emissions regulations on a state and federal level impact many of these operations as well.

Most natural gas production requires processing to remove traces of other hydrocarbons and impurities from the natural gas stream. The recovery of natural gas liquids such as propane, butane, pentanes, higher molecular weight hydrocarbons and other condensates is a value-adding process throughout much of the gas processing industry. Other trace products such as hydrogen sulfide (H_2S) and carbon dioxide (CO_2) are referred to as acid gas and must be removed from the gas stream to prevent corrosion of pipelines and equipment for safety reasons.

Natural gas has been referred to as a low-carbon fuel, as its combustion produces significantly less carbon dioxide emissions than coal and petroleum-based fuels. However, to understand the implications for climate change, one must look at not only the GHG emissions from combustion in a vehicle or power plant but also those from production activities. For natural gas, the primary concern is leakage and venting throughout the supply chain, as methane (CH_4), a potent GHG, is its primary constituent.

The leakage of methane, the main component of natural gas, into the atmosphere also raises environmental concerns. The Energy Information Administration state that methane emissions from all sources account for about 1% of total US GHG emissions, but about 9% of the GHG emissions based on global warming potential. Methane can leak at any stage of the entire process leading to consumption.

Normally, field production, gathering and cleaning, separation of water or oil from associated gas, and the extraction of natural gas liquids reduce gross natural gas production by about 6–10%. In addition, transmission and distribution consume another 3–8%, further reducing the gross natural gas volume. As a result, only about 85–90% of the gross production in the United States reaches end users. However, whether the natural gas flows from a vertical well or from a horizontal well, the process leading to consumption does not vary.

A further examination of the carbon footprint on a *per-well* basis may generate an apparent contradictory result. The carbon footprint of a horizontal well far exceeds that of a typical vertical well since the drilling process, the completion process, and the production stimulation process (hydraulic fracturing) require more carbon-based fuels, more drilling mud, and more water. Further, running the required equipment and pumps produces more emissions.

In 2011, the US Environmental Protection Agency doubled its estimates of methane leakage for the US natural gas industry, in part because of the inclusion of emissions from shale gas production for the first time. One key activity that can produce significant methane emissions is shale gas well completions. When flowback water is removed from the well prior to the beginning of gas production, natural gas can be vented to the atmosphere over the course of several days. Periodically, a shale gas well may need a *work over* to improve gas flow, which can involve hydraulically fracturing the well again, and thus further methane emissions can occur if these operations are not controlled (Osborn et al., 2011).

In reality, natural gas operators often take steps to limit these emissions. The EPA's Natural Gas STAR program, an industry and government partnership to reduce methane emissions, has been reporting significant (approximately 50%) emission reductions through the use of flaring and *reduced emissions completions* (RECs), which allow them to capture gas that otherwise would have been vented to the atmosphere (Burnham et al., 2012). However, the estimates of savings lack transparency, as they are highly aggregated to protect confidential business information. Another area of uncertainty when estimating the impacts of these emissions is projecting future well productivity, which is an important factor in life-cycle calculations (Berman, 2012; Branosky et al., 2012). Because shale gas production is so new, these projections range widely, and if wells are less productive than the industry projects, then the emissions impacts of well completions will be of greater importance.

Several studies have been released that have estimated the life-cycle GHG emissions of shale gas; however, results have varied due to differences in methodology and data assumptions (Burnham et al., 2012; Howarth et al., 2011a,b; Weber and Clavin, 2012). The EPA does not explicitly examine shale gas leakage, but, rather, examines the entire natural gas industry; however, previous EPA estimates for natural gas leakage prior to large-scale shale gas production were 1.4% for the life cycle and 0.4% for the production phase (Kirchgessner et al., 1997). While the estimated leakage rate has increased significantly from previous estimates for various activities associated with production, those for other stages such as transmission and distribution have declined due to replacement of older pipelines, thereby reducing the overall impact. On the other hand, Cornell researchers estimated a base case leakage rate for shale gas of 5.8% for the life cycle; however, they do not account for technologies that capture vented methane and do not include several data points that likely overestimate emissions.

Using current leakage estimates for large-scale production, natural gas methane emissions account for approximately 15% of the total life-cycle GHG emissions on a 100-year timescale, and the relative benefits of natural gas depend on how it is ultimately used. For example, most studies show that natural gas power plants can provide approximately 30–50% reduction in GHG emissions, depending on the plant's efficiency, as compared to a typical coal plant (Burnham et al., 2012). For light-duty vehicles, use of compressed natural gas may provide nearly a 10% reduction in GHG emissions as compared to gasoline (Burnham et al., 2012). However, for heavy-duty natural gas vehicles using spark-ignited engines, such as a transit bus, there may be no GHG benefit as compared to diesel vehicles, owing to the efficiency advantage of compression–ignition engines.

Another local air pollutant of growing concern is crystalline silica dust, which can be generated from the sand proppant. Silica dust can be generated in the mining and transporting of sand to the well site and in the process of moving and mixing sand into the hydraulic fracturing fluid on the well pad. Crystalline silica dust within the respirable size range ($<4\,\mu m$) is considered a hazardous air pollutant and a carcinogen. In addition to an increased risk of lung cancer, exposure to crystalline silica can lead to a chronic, inflammatory lung disease called silicosis.

5.3.2 Water Contamination

In the United States, shale basins are found across most of the lower 48 states (see Chapter 2). Currently, the most active shale basins in the United States are (alphabetically) the Antrim Shale, the Barnett Shale, the Fayetteville Shale, the Haynesville/Bossier Shale, the Marcellus Shale, and New Albany Shale, with other gas shale formation under serious investigation as gas producers (see Chapter 2) (EIA, 2011a). The only commonality is the name *shale* and, as is the case with processing the natural gas, each gas shale basin presents its own unique set of challenges with respect to water resource management.

In terms of water resources, the issue is the application of hydraulic fracturing to release the gas form the shale formation—high-volume hydraulic fracturing to create fissures in the rock to release gas or oil trapped inside. Thus, water-related issues in shale drilling are leading to growing and complex policy and regulatory challenges and environmental compliance hurdles that could potentially challenge shale gas production expansion and increase operational costs.

Water consumption for hydraulic fracturing occurs during (i) drilling, (ii) extraction and processing of proppant sands, (iii) testing natural gas transportation pipelines, and (iv) operation of gas processing plants. Typically, for most shale basins, water is acquired from local water supplies, including (i) surface water bodies, such as rivers, lakes, and ponds, (ii) groundwater aquifers, (iii) municipal water supplies, (iv) treated wastewater from municipal and industrial treatment facilities, and (v) produced and/or flowback water that is recovered, treated, and reused. In regions where hydraulic fracturing occurs, the sources of water should be well documented.

Water that originates from hydraulic fracturing often contains chemical additives to help carry the proppant and may become enriched in salts after being injected into shale formations. Therefore, the water that is recovered during natural gas production must be either treated or disposed of in a safe manner—typically by injection into deep, highly saline formations through one or more wells drilled specifically for that purpose and by following clearly defined regulations. Flowback water is infrequently reused in hydraulic fracturing because of the potential for corrosion or scaling, where the dissolved salts may precipitate out of the water and clog parts of the well or the formation.

In addition to fracturing fluid additives, wastewater from shale gas extraction may contain high levels of total dissolved solids (TDS) metals and naturally occurring radioactive materials. Furthermore, the amount of saline formation water produced from gas shale formations varies widely—from zero to several hundred barrels per day. The water comes from the gas shale itself or from adjacent formations that are connected through the fracture-induced fracture network. The water, like flowback water, is normally highly saline and must be treated and/ or disposed of, typically by injection into deep saline formations, which is also subject to clearly defined regulations. In fact, in some oil-and-gas-producing states, regulatory agencies have implemented regulations regarding the disclosure of chemicals used in the process of hydraulic fracturing.

Well casing provides a protective barrier from potential contamination from hydraulic fracturing fluid, oil, and natural gas flowing from the well. However some of the risks to water quality occur from ground and surface spills—such as shale gas drilling water contamination or other mishandling of wastewater—rather than from the hydraulic fracturing process itself.

Nevertheless, the potential contamination of groundwater raises another environmental concern. The hydraulic fracturing process requires the use of hundreds of thousands of gallons of water treated with chemicals that facilitate both the suspension of the proppant (sand, most frequently) and the lubrication of the conveying mediums. In the development of an entire field, the amount of water injected into a shale formation could reach into the hundreds of millions of gallons. Although field operators retrieve most of the injected water upon completion of the hydraulic fracturing stimulation, a significant quantity of water and chemicals remains within the formation.

Development of several shale formations, for example, the Barnett near Fort Worth, Texas, is occurring near major population centers. As a result, some environmentalists claim that potential leakage of chemicals used in the hydraulic fracturing process poses a health and safety risk and are calling for stricter regulation. The natural gas industry responds to the concern by pointing out that "the chemical injections [into the shale formations] are happening thousands of feet below the surface, whereas groundwater is usually just hundreds of feet deep."

Some states, including New York, have issued regulatory requirements for *responsible development* of shale formations. These regulations include guidelines for the use and disposal of water, the protection of groundwater, and the use of chemicals. Further, the regulatory requirements include (i) review of each drilling application for environmental compliance, (ii) complete environmental assessment of all proposed oil or gas wells that are within 2000 ft of a municipal water well, (iii) strict review of the well design to ensure groundwater protection, (iv) on-site of inspection of drilling operations, and (v) enforcement of strict restoration rules when drilling ends.

5.3.2.1 Water Consumption

Although water is used in several stages of the shale gas life cycle, the majority of water is typically consumed during the production stage. This is primarily due to the large volumes of water (2.3–5.5 million gallons) required to hydraulically fracture a well (Clark et al., 2011). Water in amounts of 190,000–310,000 gallons is also used to drill and cement a shale gas well during construction (Clark et al., 2011). After fracturing a well, anywhere from 5% to 20% of the original volume of the fluid will return to the surface within the first 10 days as flowback water. An additional volume of water, equivalent to anywhere from 10% to almost 300% of the injected volume, will return to the surface as produced water over the life of the well. It should be noted that there is no clear distinction between so-called flowback water and produced water, with the terms typically being defined by operators based on the timing, flow rate, or sometimes composition of the water produced.

The rate at which water returns to the surface is highly dependent on the geology of the formation. In the Marcellus play, operators recycle 95% of the flowback, whereas in the Barnett and Fayetteville plays, operators typically recycle 20% of the flowback. Water management and reuse are local issues and often depend on the quality and quantity of water and the availability and affordability of management options (Veil, 2010). Over a 30-year life cycle, assuming a typical well is hydraulically fractured three times during that time period, construction and production of shale gas typically consume between 7,090,000 and 16,810,000 gallons of water per well.

Once the gas is produced, it is processed, transported and distributed, and ultimately used. Water consumption occurs in each of these stages as well, with the most significant nonproduction consumption potentially occurring during end use. Although natural gas can be combusted directly with no additional water consumption, if the end use of the gas is a vehicle tank, it might be compressed via an electric compressor. The electricity for compression is associated with water consumption of 0.6−0.8 gal/GGE (gallon per gasoline gallon equivalent), making the total consumption for the vehicle life cycle 1.0−2.5 gal/GGE depending on location and the extent that flowback water is recycled. For comparison, vehicle life-cycle water consumption associated with the use of conventional natural gas is between 0.9 and 1.1 gal/GGE, gasoline is between 2.6 and 6.6 gal/GGE, and corn ethanol is between 26 and 359 gal/GGE (Wu et al., 2011).

5.3.2.2 Water Quality

Concerns over water quality focus on potential drinking water contamination by methane or fluids from hydraulic fracturing activities. The possible pathways for this contamination include underground leakage from the wellbore to drinking water aquifers and improper disposal or accidental leakage of hydraulic fracturing fluids to surface water bodies. Owing to the depth of most shale plays, it is unlikely that a credible pathway (independent of the wellbore) exists for fluids to flow from the fractures within the shale through thousands of feet of overlaying rock into a drinking water aquifer. However, shallower shale deposits may be vulnerable to this direct connection, as is suggested by EPA's ongoing groundwater investigation in Pavillion, Wyoming, where as little as 400 ft separated gas deposits from drinking water resources.

For deep formations, contamination may occur due to defects in the wellbore. When the annulus between the well casing and surrounding geology is not adequately sealed during well installation, methane can migrate from the shale resource up the outside of the wellbore to shallow aquifers where it could dissolve in the drinking water. Another possible pathway for contamination is a defect in the casing at a shallow depth, allowing gas to flow from inside the wellbore to the aquifer.

Faulty well construction appears to have caused one of the largest documented instances of water contamination, which occurred in Bradford County, Pennsylvania, after wells had been drilled but before any hydraulic fracturing took place. In addition to faulty well construction, uncased, abandoned wells may also provide pathways for methane migration to occur (Osborn et al., 2011). The most obvious, and perhaps most easily prevented, pathway for contamination is intentional dumping or accidental spilling of flowback water on the surface. A common cause of accidental spillage is overflows from retention ponds during major rain events.

Contaminants in flowback water from the mineral formation, such as naturally occurring radioactive materials (NORM), or from additives to the hydraulic fracturing fluid can be a health concern when present in significant concentrations. EPA's investigation into possible groundwater contamination at Dimock, Pennsylvania, was launched out of concern over such toxic substances. While there are no Federal drinking water standard limits for methane, it is nevertheless a hazard in water because at sufficient concentrations it can volatilize and collect in houses, which can lead to suffocation or serve as a fuel for fire and explosions.

5.3.2.3 Water Treatment

The treatment of wastewater is a critical issue for unconventional gas production—especially in the case of the large amounts of water customarily used for hydraulic fracturing. After being injected into the well, part of the fracturing fluid (which is often almost entirely water) is returned as flowback in the days and weeks that follow. The total amount of fluid returned depends on the geology; for shale it can run from 20% to 50% of the input, the rest remaining bound to the clays in the shale rock. Flowback water contains some of the chemicals used in the hydraulic fracturing process, together with metals, minerals, and hydrocarbons leached from the reservoir rock. High levels of salinity are quite common and, in some reservoirs, the leached minerals can be weakly radioactive, requiring specific precautions at the surface. Flowback returns (like wastewater from drilling) require secure storage on-site, preferably fully contained in stable, weather-proof storage facilities as they do pose a potential threat to the local environment unless handled properly.

Once separated out, there are different options available for dealing with wastewater from hydraulic fracturing. The optimal solution is to recycle it for future use, and technologies are available to do this, although they do not always provide water ready for reuse for hydraulic fracturing on a cost-effective basis. A second option is to treat wastewater at local industrial waste facilities capable of extracting the water and bringing it to a sufficient standard to enable it to be either discharged into local rivers or used in agriculture. Alternatively, where suitable geology exists, wastewater can be injected into deep rock layers.

5.3.2.4 Water Recycling

The issue of shale gas regulation is dominated by hydraulic fracturing, the key feature of shale gas that separates it from well-regulated conventional gas production. However, existing regulations to protect water resources during oil and gas development are also affected by the greater intensity of water, energy, and infrastructure used in shale gas operations.

This consequence is driving significant uncertainty in the United States, which is still adapting to the new industry. The speed of industry growth has outpaced the availability of rigorous data on its potential impact, which has hindered the ability of government to adequately assess and regulate operations. To resolve this issue, there has been renewed focus by the US federal government on establishing better understanding of the potential impacts of shale gas development to most effectively regulate this critical new energy resource.

A large volume of water is needed for the development of shale gas plays. Water is used for drilling, where it is mixed with clay minerals to form drilling mud. This mud is used to cool and lubricate the drill bit, provide wellbore stability, and also carry rock cuttings to the surface.

Water is also used in significant volumes in hydraulic fracturing. In addition to water and sand, a small concentration of other additives is added to the fluid to improve fracking efficiency. Chesapeake Energy cites a figure of 4.5 million gallons of fluid for the fracturing of a typical horizontal well. This significant volume of water needs a plentiful source. In the United States, the water is typically trucked to the drilling location or transported via temporary pipelines.

A typical fracking fluid is more than 98% v/v water and sand. The other 2% consists of additives which may vary depending on the particular well and operator. Typically, additives include many substances that are commonly found in small measure in various household products.

During a typical hydraulic fracturing process, the fracking fluid is transmitted down a cased wellbore to the target zones and then forced deep into the targeted shale gas formations. In order to minimize the risk of any groundwater contamination, good drilling practice normally requires that one or more strings of steel casing are inserted into the well and cemented into place so as to ensure that the entire wellbore, other than the production zone, is completely isolated from the surrounding formations including aquifers.

Most oil or gas-bearing shale in the United States tends to be at least 4600 ft below the surface, whereas aquifers are generally no more than 1550 ft below the surface. Given the thickness of rock separating target shale formations from overlying aquifers, and the extremely low permeability of shale formations themselves, and also assuming the implementation of good oilfield practices (such as casing and cementing), it is considered by the industry that the risk of contamination of overlying aquifers as a result of hydraulic fracturing operations is remote. Instances where contamination of aquifers has been alleged are generally believed to have involved poor drilling practices—in particular, poor casing and cementing of a well or poor construction of surface storage facilities.

Currently, most of the flowback fluid from fracking operations is either transported from well sites for disposal or is processed for reuse in further operations. Suspended solids must be removed from the water before reuse. Recycling this water can be costly and is a major focal point of many environmental groups and environmental regulators. New, more efficient, technologies have been developed which allow fracking fluid to be recycled on-site at reduced cost.

However, hydraulic fracturing does not require water that is of potable (drinking water) quality. Recycling wastewater helps conserve water use and provide cost-saving opportunities. In gas recovery from the Marcellus shale, there are examples of companies reusing up to 96% of the produced water. Other examples of recycling and reuse include (KPMG, 2012) the following:

1. The use of portable distilling plants to recycle water in the Barnett Shale, particularly in regions such as the Granite Wash field in North Texas, where water resources are more critical than in other shale basins in the United States.
2. A water purification treatment center can recycle several thousand barrels of flowback and produced water per day generated from extracting oil and natural gas from a shale formation—this approach is being used in the Eagle Ford Shale and in the Marcellus Shale.
3. The Marcellus Shale also employs vapor recompression technology to reduce the cost of recycling fracturing water by using waste heat. The unit produces water vapor and solid residue that is disposed of in a waste facility. In addition, to reduce contamination risks during shale operations, many gas companies in the Marcellus Shale are reducing the amount of chemical additives used in fracturing fluid while producing shale gas.
4. A wastewater treatment company specializing in the oil and gas industry has designed a mobile integrated treatment system for hydraulic fracturing that allows the reuse of water for future drilling. Using dissolved air flotation technology, the system can treat up to 900 gal/min of fracking flowback water. The accelerated water treatment reduces the equipment burdens and logistics of traditional treatment methods and could significantly reduce operational costs.
5. Produced water can have high TDS concentrations that can be difficult to treat. Thermal distillation, reverse osmosis (RO), and other membrane-based desalination technologies can be deployed to desalinate produced water to a level fit for purpose.

Fluids other than water may be used in fracking processes, including carbon dioxide, nitrogen, or propane, although their use is currently much less widespread than water.

5.3.2.5 Water Disposal

When water returns to the surface from a shale drilling operation, it may be disposed of in a variety of ways, depending on the shale basin: (i) reused in a new well, with or without treatment, (ii) injected into on- or off-site disposal wells regulated by the US Environmental Protection Agency, (iii) taken to a municipal wastewater treatment plant or a commercial industrial wastewater treatment facility—most wastewater treatment plants are not capable of treating the contaminants in shale gas wastewater, or (iv) discharged to a nearby surface water body.

In the Marcellus Shale, one of the largest shale basins in the United States located in Pennsylvania and New York state, a large proportion of the hydraulic fracturing fluid is usually recovered after drilling and stored on-site in evaporation pits. Recovered fluid may be trucked off-site for use in another fracking operation or for treatment and disposal in surface waters, underground reservoirs, or at a wastewater treatment facility. The remainder of the fluid remains underground (Veil, 2010). However, in the water-deprived shale basins of Texas (such as Eagle Ford), more of the hydraulic fracturing fluid may remain underground. This water is much harder to track than surface water, which may lead to increased short- and long-term risks for shale gas companies.

5.3.3 Fluids Management

A variety of waste fluids are generated on-site at shale gas wells. During drilling, used mud and saturated cuttings are produced and must be managed. The volume of mud approximately correlates with the size of the well drilled, so a horizontal Marcellus well may generate twice as much drilling waste as a single vertical well; however, as discussed above, it will replace four such holes (Arthur et al., 2008). Drilling wastes can be managed on-site either in pits or in steel tanks. Each pit is designed to keep liquids from infiltrating vulnerable water resources. On-site pits are a standard in the oil and gas industry but are not appropriate everywhere; they can be large, and they disturb the land for an extended period of time. Steel tanks may be required to store drilling mud in some environments to minimize the size of the well site *footprint* or to provide extra protection for a sensitive environment. Steel tanks are not, of course, appropriate in every setting either, but in rural areas or pits or ponds, where space is available at the well site, steel tanks are usually not needed (Arthur et al., 2008).

Horizontal drilling development has the power to reduce the number of well sites and to group them so that management facilities such as storage ponds can be used for several wells. Makeup water is used throughout the development process to drill the well and to form the basis of the hydraulic fracturing fluid. Large volumes of water may be needed and are often stored at the well site in pits or tanks. For example, surface water can be piped into the pit during high-water run-off periods and used during the year for drilling and fracture treatments in nearby wells. Storage ponds are not suitable everywhere in the area of a shale gas resource—just as steel tanks are appropriate in some locations but not in others.

5.3.4 Induced Seismic Activity

Disposal of flowback water from hydraulic fracturing depends on the availability of suitable injection wells. For example, the limited availability of suitable geology in Pennsylvania has led to hauling flowback water to Ohio for injection. The increased injection activity has been linked to *seismic events* or earthquakes. Additional studies have indicated that injection activities in Arkansas have been linked to nearby earthquakes (Horton, 2012).

A properly located injection well will not cause earthquakes. A number of factors must be present to induce seismic events at a disposal site. In order for earthquakes to occur, a fault must exist nearby and be in a near-failure state of stress. The injection well must have a path of communication to the fault, and the fluid flow rate in the well must be at a sufficient quantity and pressure for a long enough time to cause failure along the fault or system of faults. A recent National Research Council study concludes that the majority of disposal wells for hydraulic fracturing wastewater do not pose a hazard for induced seismicity. This report also concludes that the process of hydraulic fracturing itself does not pose a high risk for inducing felt seismic events (NRC, 2012).

Nevertheless, there have been instances of earthquakes associated with unconventional gas production, for example, the case of the Cuadrilla Shale gas operations near Blackpool in the United Kingdom, or a case near Youngstown, Ohio, in the United States, which has been provisionally linked to injection of wastewater, an operation that is similar in some respects to hydraulic fracturing. The registered earthquakes were small, of a magnitude of around 2 on the Richter scale, meaning they were discernible by humans but did not create any surface damage.

Because it creates cracks in rocks deep beneath the surface, hydraulic fracturing always generates small seismic events; these are actually used by petroleum engineers to monitor the process. In general, such events are several orders of magnitude too small to be detected at the surface: special observation wells and very sensitive instruments need to be used to monitor the process. Larger seismic events can be generated when the well or the fractures happen to intersect, and reactivate, an existing fault. This appears to be what happened in the Cuadrilla case.

Hydraulic fracturing is not the only anthropogenic process that can trigger small earthquakes. Any activity that creates underground stresses carries such a risk. Examples linked to construction of large buildings, or dams, have been reported. Geothermal wells in which cold water is circulated underground have been known to create enough thermally induced stresses to generate earthquakes that can be sensed by humans (Cuenot et al., 2011) and the same applies to deep mining (Redmayne et al., 1998).

In order to circumvent any such issues arising from hydraulic fracturing, it is essential for unconventional gas development engineers to make a careful survey of the geology (with the geologists) of the area to assess whether deep faults or other geological features present an enhanced risk and to avoid such areas for fracturing. In any case, multidisciplinary monitoring is necessary so that operations can be suspended and corrective actions taken if there are signs of increased seismic activity.

5.4 REMEDIATION REQUIREMENTS AND OUTLOOK

The occurrence and production of natural gas from fractured, organic-rich Paleozoic and Mesozoic shale formations in the United States may be better understood by considering source rock, reservoir, seal, trap, and generation migration processes within the framework of a petroleum system. The system concept must be modified, however, inasmuch as organic shale formations are both source and reservoir rocks and, at times, seals. Additional consideration must be given to the origin of the gas, whether biogenic or thermogenic, in defining the critical moment in the evolution of potentially producible hydrocarbons.

This emerging resource can be considered a technology-driven play as achieving gas production out of otherwise unproductive rock requires technology-intensive processes. Maximizing gas recoveries requires far more wells than would be the case in conventional natural gas operations.

Remediation requirements become of greater importance as wells reach the end of their life cycles. More than half of the total production of a well is usually achieved in the first 10 years of operative well life. When a well can no longer produce shale gas economically, it is plugged and abandoned according to the standards of each state. Disturbed areas, such as well sites and access roads, are reclaimed back to the native vegetation and contours, or to conditions specified by the landowner.

Improperly closed or abandoned shale gas wells may create human health and safety risks, as well as air pollution and surface and groundwater contamination risks. Most states require operators to post a bond or some form of financial security to ensure compliance, and also to ensure there are funds to properly plug the well once production ceases. However, the size of the bond may cover only a small fraction of the site reclamation costs.

The economics of shale gas development encourages the transfer of assets from large entities to smaller ones. With the assets go the liabilities, but without a mechanism to prevent the new owners from assuming reclamation liabilities beyond their means, the economics favor default on well-plugging and site restoration obligations.

In fact, a combination of improved technology and shale-specific experience has also led—and will continue to lead—to improvements in recovery factors and reductions in decline rates. It is now recognized that each gas shale resource requires a specific completion technique, which can be determined through careful analysis of rock properties. Continuous efforts to make the correct selection of well orientation, stimulation equipment, fracture size, and fracking fluids will serve to enhance the performance of a well and the overall recovery of gas.

Indeed, for developed shale formations in North America, the combined benefits of improved technology and increased experience will continue to provide enhanced production over time. Both the expected ultimate recovery per well and the peak production per well will continue to increase as developed shale gas formations move to maturity.

Following on from the advances in shale gas in the United States, a number of oil and gas companies will be willing to apply the techniques developed in North America in new geological basins and markets out-side North America. A considerable number of regions around the world have been the focus of interest for their shale potential—in fact, 48 major shale basins are identified in 32 countries around the world that are pro-spects for development (EIA, 2011b).

These prospects include a number of shale formations across Europe where organic-rich shale sediments are present, including (i) Lower Paleozoic shale formations, which extend from Eastern Denmark and Southern Sweden to Northern and Eastern Poland, (ii) Carboniferous shale formations, which extend from North West England through the Netherlands and northwest Germany to southwest Poland, and (iii) Lower Jurassic bituminous shale formations, which extend from the South of England to the Paris Basin in France, the Netherlands, Northern Germany, and Switzerland. Poland and France are identified (EIA, 2011b) as countries with some of the largest estimated shale gas technically recoverable resources in Europe—both countries are currently highly dependent on imported gas to meet domestic demand.

Furthermore, horizontal wells with horizontal legs up to one mile or more in length are widely used to access the reservoir to the greatest extent possible. Multistage hydraulic fracturing, where the shale is cracked under high pressures at several places along the horizontal sec-tion of the well, is used to create conduits through which gas can flow. Microseismic imaging allows operators to visualize where this fracture growth is occurring in the reservoir.

Although fracture and matrix permeability, enhanced by applica-tion of appropriate well stimulation treatments, are key to achieving economical gas flow rates, sufficient amounts of organic matter (either for generation of thermogenic gas or as a microbial feedstock) must initially have been present to have generated the reservoir gas. Therefore, deciphering the thermal history of the organic matter within the shales and analyzing the rock mechanics response of the shale matrix and organic matter to local and regional stresses are critical steps in establishing their complex relationship to gas producibility. The poor quality of one factor (e.g., low adsorbed gas) may be com-pensated for by another factor (e.g., increased reservoir thickness); however, shale gas production cannot always be achieved even where optimum combinations of geological and geochemical factors appar-ently are present.

However, as a technology-driven resource, the rate of development of shale gas may become limited by the availability of required resources, such as freshwater, fracture proppant, or drilling rigs capable of drilling wells miles in length.

Thus, important challenges for developing the shale gas resources are (i) the significant depth and (ii) the lack of information for many of the resources.

In areas where the resources are present, companies must continue to focus on the environmental development before setting their sights on a deeper target with an uncertain payoff.

On the other hand, in areas where the shale gas development has already occurred and new resources are discovered and opened up to development, there may be an infrastructure advantage. Drilling pads, roadways, pipelines, gathering systems, surveying work, permit preparation data, and landowner relationships might still be useful for developing future shale resources.

There is potential for a heavy draw on freshwater resources because of the large quantities required for hydraulic fracturing fluid. The land-use footprint of shale gas development is not expected to be much more than the footprint of conventional operations, despite higher well densities, because advances in horizontal drilling technology allow for up to 10 or more wells to be drilled and produced from the same well site.

Finally, there is potential for a high carbon footprint through emissions of carbon dioxide (CO_2), a natural impurity in some shale gas.

REFERENCES

Arthur, J.D., Langhus, B., Alleman, D., 2008. An Overview of Modern Shale Gas Development in the United States. ALL Consulting, Tulsa, OK, <http://www.all-llc.com/publicdownloads/ALLShaleOverviewFINAL.pdf>.

Arthur, J.D., Bohm, B., Cornue, D., 2009. Environmental considerations of modern shale developments. Paper No. SPE 122931. Proceedings of the SPE Annual Technical Meeting, October 4–7, New Orleans, LA.

Berman, A.E., 2012. After the gold rush: a perspective on the future US natural gas supply and price. Proceedings of the Association for the Study of Peak Oil and Gas, Vienna.

Branosky, E., Stevens, A., Forbes, S., 2012. Defining the Shale Gas Life Cycle: A Framework for Identifying and Mitigating Environmental Impacts. World Resources Institute, Washington, DC (Working Paper).

Burnham, A., Han, J., Clark, C., Wang, M., Dunn, J., Palou-Rivera, I., 2012. Life-cycle greenhouse gas emissions of shale gas, natural gas, coal, and petroleum. Environ. Sci. Technol. 46 (2), 619–627.

Clark, C., Han, J., Burnham, A., Dunn, J., Wang, M., 2011. Life-Cycle Analysis of Shale Gas and Natural Gas. Argonne National Laboratory, Argonne, IL (Report No. ANL/ESD/11-11).

Clark, C., Burnham, A., Harto, C., Horner, R., 2012. Hydraulic Fracturing and Shale Gas Production: Technology, Impacts, and Policy. Argonne National Laboratory, Argonne, IL, September 10.

Cuenot, N., Frogneux, M., Dorbath, C., Calo, M., 2011. Induced microseismic activity during recent circulation tests at the EGS site of Soultz-sous-Forêts (France). Proceedings of the 36th Workshop on Geothermal Reservoir Engineering, Stanford, CA.

EIA, 2011a. Shale Gas and Shale Oil Plays. Energy Information Administration. United States Department of Energy, Washington, DC, July, <www.eia.gov> (accessed 29.03.13).

EIA, 2011b. World Shale Gas Resources: An Initial Assessment of 14 Regions Outside the United States. Energy Information Administration, United States Department of Energy, <www.eia.gov> (accessed 29.03.13).

GAO, 2012. Information on shale resources, development, and environmental and public health risks. Report No. GAO-12-732. Report to Congressional Requesters. United States Government Accountability Office, Washington, DC, September.

Gaudlip, A.W., Paugh, L.O., Hayes, T.D., 2008. Marcellus water management challenges in Pennsylvania. Paper No. SPE 119898. Proceedings of the Shale Gas Production Conference, November 16–18, Fort Worth, TX.

Horton, S., 2012. Disposal of hydrofracking waste fluid by injection into subsurface aquifers triggers earthquake swarm in central Arkansas with potential for damaging earthquake. Seismol. Res. Lett. 83 (2), 250–260.

Howarth, R.W., Santoro, R., Ingraffea, A., 2011a. Methane and the greenhouse-gas footprint of natural gas from shale formations. Clim. Change 106, 679–690.

Howarth, R.W., Santoro, R., Ingraffea, A., 2011b. Methane and the greenhouse-gas imprint of natural gas from shale formations. Clim. Change 106 (4), 1–12.

King, C.W., Webber, M.E., 2008. Water intensity of transportation. Environ. Sci. Technol. 42 (21), 7866–7872.

Kirchgessner, D.A., Lott, R.A., Cowgill, R.M., Harrison, M.R., Shires, T.M., 1997. Estimate of methane emissions from the US natural gas industry. Chemosphere 35, 1365–1390.

KPMG, 2012. Watered-Down: Minimizing Water Risks in Shale Gas and Oil Drilling. KPMG Global Energy Institute, KPMG International, Houston, TX.

NRC, 2012. Induced Seismicity Potential in Energy Technologies. National Research Council, The National Academies Press, Washington, DC.

Osborn, S.G., Vengosh, A., Warner, N.R., Jackson, R.B., 2011. Methane contamination of drinking water accompanying gas-well drilling and hydraulic fracturing. Proc. Natl. Acad. Sci. 108 (20), 8172–8176.

O'Sullivan, F., Paltsev, S., 2012. Shale gas production: potential versus actual greenhouse gas emissions. Environ. Res. Lett. 7, 1–6.

Redmayne, D.W., Richards, J.A., Wild, P.W., 1998. Mining-Induced earthquakes monitored during pit closure in the Midlothian Coalfield. Q. J. Eng. Geol. Hydrogeol. Geol. Soc. London, UK 31 (1), 21.

Schrag, D.P., 2012. Is shale gas good for climate change? Dædalus, J. Am. Acad. Arts Sci. 141 (2), 72–80.

Shine, K.P., 2009. The global warming potential—the need for an interdisciplinary retrial. Clim. Change 96 (4), 467–472.

Spellman, F.R., 2013. Environmental Impacts of Hydraulic Fracturing. CRC Press, Taylor & Francis Group, Boca Raton, FL.

Stephenson, T., Valle, J.E., Riera-Palou, X., 2011. Modeling the relative GHG emissions of conventional and Shale gas production. Environ. Sci. Technol. 45, 10757–10764.

US EPA, 2012. Regulation of Hydraulic Fracturing Under the Safe Drinking Water Act. United States Environmental Protection Agency, Washington, DC, <water.epa.gov>.

Veil, J.A., 2010. Water Management Technologies Used by Marcellus Shale Gas Producers. Argonne National Laboratory, Argonne, IL (Report No. ANL/EVR/R-10/3).

Weber, C.L., Clavin, C., 2012. Life cycle carbon footprint of Shale gas: review of evidence and implications. Environ. Sci. Technol. 46, 5688–5695.

Wu, M., Mintz, M., Wang, M., Arora, S., Chiu, Y., 2011. Consumptive Water Use in the Production of Ethanol and Petroleum Gasoline—2011 Update. Argonne National Laboratory, Argonne, IL (Report No. ANL/ESD/09-1).

NATURAL GAS CONVERSION TABLE[*]

1 cubic foot (cf) = 1000 Btu
1 Ccf = 100 cubic feet = 1 therm = 100,000 Btu
1 Mcf = 1000 cubic feet = 1 MM Btu
1 MMcf = 1 million cubic feet = 10,000 MM Btu
1 Bcf = 1 billion cubic feet = 1×10^9 ft^3 = 1 million MM Btu
1 Tcf = 1 trillion cubic feet = 1×10^{12} ft^3

[*]Based upon an approximate natural gas heating value of 1000 Btu per cubic foot.

Ft³	Ccf	Mcf	MMcf	Therm	Dekatherm	Btu	MMBtu	kJ	kWh
1	0.01	0.001	0.000001	0.01	0.001	1000	0.001	1054	0.293
100	1	0.1	0.0001	1	0.1	100,000	0.1	105,461.5	29
1000	10	1	0.001	10	1	1,000,000	1	1,054,615	293
1,000,000	10,000	1000	1	10,000	1000	1.00E+09	1000	1,054,615,000	293,071
100	1	0.1	0.0001	1	0.1	100,000	0.1	105,500	29
1000	10	1	0.001	10	1	1,000,000	1	1,054,615	293
0.001	0.00001	0.000001	0.000000001	0.00001	0.0001	1	0.000001	1.055	0
1000	10	1	0.001	10	1	1,000,000	1	1,054,615	293
0.0009	0.00001	0.000001	0	0.00001	0.0001	0.9482	0.000001	1	0
3.345	0.033	0.003	0.000003	0.034	0.003	3412	0.003	3600	1

GLOSSARY

Abandonment pressure A direct function of the economic premises, the static bottom pressure at which the revenues obtained from the sales of the hydrocarbons produced are equal to the well's operation costs.

Abiogenic gas Gas formed by inorganic means.

Absolute permeability Ability of a rock to conduct a fluid when only one fluid is present in the pores of the rock.

Absorber See Absorption tower.

Absorption The process by which the gas is distributed throughout an absorbent (liquid); depends only on physical solubility and may include chemical reactions in the liquid phase (*chemisorption*).

Absorption oil Oil used to separate the heavier components from a vapor mixture by absorption of the heavier components during intimate contacting of the oil and vapor; used to recover natural gasoline from wet gas.

Absorption plant A plant for recovering the condensable portion of natural or refinery gas, by absorbing the higher boiling hydrocarbons in an absorption oil, followed by separation and fractionation of the absorbed material.

Absorption tower A tower or column which promotes contact between a rising gas and a falling liquid so that part of the gas may be dissolved in the liquid.

Acid deposition (acid rain) This occurs when sulfur dioxide (SO_2) and, to a lesser extent, NO_x emissions are transformed in the atmosphere and return to the earth as dry deposition or in rain, fog, or snow.

Acid gas Carbon dioxide and hydrogen sulfide; see also Sour gas.

Acoustic log See Sonic log.

Adsorption Transfer of a substance from a solution to the surface of a solid resulting in relatively high concentration of the substance at the place of contact; molecular bonding of a gas to the surface of a solid. In the case of shale, natural gas is adsorbed or bonded to the organic material in the shale.

Air pollution The discharge of toxic gases and particulate matter introduced into the atmosphere, principally as a result of human activity.

Air quality A measure of the amount of pollutants emitted into the atmosphere and the dispersion potential of an area to dilute those pollutants.

American Society for Testing and Materials (ASTM) The official organization in the United States for designing standard tests for petroleum and other industrial products.

Amine washing A method of gas cleaning whereby acidic impurities, such as hydrogen sulfide and carbon dioxide, are removed from the gas stream by washing with an amine (usually an alkanolamine).

Annulus The space between the casing and the wellbore or surrounding rock.

Anticline An area of the earth's crust where folding has made a dome-like shape in the once flat rock layers. Anticlines often provide an environment where natural gas can become trapped beneath the Earth's surface and extracted.

Aquifer An underground layer of water-bearing rock or gravel, sand or silt; the subsurface layer of rock or unconsolidated material that allows water to flow within it; aquifers can act as sources for groundwater, both usable fresh water and unusable salty water.

Associated gas Natural gas that is in contact with and/or dissolved in the crude oil of the reservoir. It may be classified as gas cap (free gas) or gas in solution (dissolved gas).

Associated gas in solution (or dissolved gas) Natural gas dissolved in the crude oil of the reservoir under the prevailing pressure and temperature conditions.

BACT Best available control technology.

Baghouse A filter system for the removal of particulate matter from gas streams, so-called because of the similarity of the filters to coal bags.

Barrel   The unit of measurement of liquids in the petroleum industry, equivalent to 42 US standard gallons or 33.6 imperial gallons.

Basin   A geological receptacle in which a sedimentary column is deposited that shares a common tectonic history at various stratigraphic levels; a closed geologic structure in which the beds dip toward a center location; the youngest rocks are at the center of a basin and are partly or completely ringed by progressively older rocks.

Bbl   See Barrel.

Bcf (billion cubic feet)   Gas measurement approximately equal to 1 trillion (1,000,000,000,000) British thermal units.

Biocide   An additive used in hydraulic fracturing fluids (and often drilling muds) to kill bacteria that could otherwise reduce permeability and fluid flow.

Biogenic   Describes material derived from bacterial or vegetation sources.

Biogenic gas   Natural gas produced by living organisms or biological processes.

Biomass   Biological organic matter.

Borehole   A generalized term for a shaft bored into the ground.

Bottom simulating reflector (BSR)   A seismic reflection at the sediment to clathrate stability zone interface caused by the different density between normal sediments and sediments laced with clathrates.

Btu (British thermal unit)   The energy required to raise the temperature of one pound of water one degree Fahrenheit.

BTU   See Btu (British thermal unit).

C_1, C_2, C_3, C_4, C_5 fractions   A common way of representing natural gas fractions containing a preponderance of hydrocarbons having 1, 2, 3, 4, or 5 carbon atoms, respectively, and without reference to hydrocarbon type.

CAA (Clean Air Act)   An act that is the foundation of air regulations in the United States.

Carbonate washing   A process using a mild alkali (e.g., potassium carbonate) process for emission control by the removal of acid gases from gas streams.

Casing   Steel pipe inserted into a wellbore and cemented into place; also used to protect freshwater aquifers or otherwise isolate a zone and serves to isolate fluids, such as water, gas, and oil, from the surrounding geologic formations.

CFR (Code of Federal Regulations)   Also known as Title 40 (40 CFR), containing the regulations for protection of the environment.

Class II injection well   A well that injects fluids into a formation rather than producing fluids. A class II injection well is a well associated with oil or natural gas production. Such wells include enhanced recovery wells, disposal wells, and hydrocarbon storage wells.

Clastic   Rock composed of pieces of pre-existing rock.

Clay   Silicate minerals that also usually contain aluminum and have particle sizes <0.002 microns, used in separation methods as an adsorbent and in refining as a catalyst.

Clean Air Act Amendments of 1990   Legislation to improve the quality of the atmosphere and curb acid rain, promoting the use of cleaner fuels in vehicles and stationary sources.

Carbonate rock   A rock consisting primarily of a carbonate mineral, such as calcite or dolomite, the chief minerals in limestone and dolostone, respectively.

Casing   Used to line the walls of a gas well to prevent collapse of the well, and also to protect the surrounding earth and rock layers from being contaminated by petroleum or the drilling fluids.

Christmas tree   The series of pipes and valves that sit on top of a producing gas well, used in place of a pump to extract the gas from the well.

Coal Bed Methane (coalbed methane)   Methane from coal seams, released or produced from the seams when the water pressure within the seam is reduced by pumping from either vertical or inclined to horizontal surface holes.

Completion   Includes the steps required to drill and assemble casing, tubes, and equipment to efficiently produce oil or gas from a well. For shale gas wells, this includes hydraulic fracturing activities.

Composition   The make-up of a gaseous stream.

Compression   Reduction in volume of natural gas compressed during transportation and storage.

Condensate   A mixture of light hydrocarbon liquids obtained by condensation of hydrocarbon vapors: predominantly butane, propane, and pentane with some heavier hydrocarbons and relatively little methane or ethane; see also Natural gas liquids.

Conductor casing   Prevents collapse of the loose soil near the surface of a borehole.

Contingent resource   The amounts of hydrocarbons estimated at a given date that are potentially recoverable from known accumulations but are not considered commercially recoverable under the economic evaluation conditions corresponding to that date.

Conventional limit   The reservoir limit established according to the degree of knowledge of (or research into) the geological, geophysical, or engineering data available.

Core   A cylindrical rock sample taken from a formation when drilling, used to determine the rock's permeability, porosity, hydrocarbon saturation, and other productivity-associated properties.

Corrosion inhibitor   A chemical compound that decreases the corrosion rate of a metal or an alloy.

Cryogenic plant   A processing plant capable of producing liquid natural gas products, including ethane, at very low operating temperatures.

Cryogenic process   A process involving low temperatures.

Cubic foot   A unit of measurement for volume, an area 1 foot long, by 1 foot wide, by 1 foot deep.

Cutting   A piece of rock or dirt that is brought to the surface of a drilling site as debris from the bottom of a well, often used to obtain data for logging.

Cyclone   A device for extracting dust from industrial waste gases. It is in the form of an inverted cone into which the contaminated gas enters tangentially from the top; the gas is propelled down a helical pathway, and the dust particles are deposited by means of centrifugal force onto the wall of the scrubber.

Darcy's law   Describes the flow of liquid through a porous medium.

Debutanization   Distillation to separate butane and lighter components from higher boiling components.

Decline rate   The rate at which the production rate of a well decreases.

De-ethanization   Distillation to separate ethane and lighter components from propane and higher boiling components, also called de-ethanation.

Dehydration   Water removal; in the present context, from natural gas streams.

Demethanization   The process of distillation in which methane is separated from the higher boiling components, also called demethanation.

Depentanizer   A fractionating column for the removal of pentane and lighter fractions from a mixture of hydrocarbons.

Depleted reservoirs   Reservoirs that have already been tapped of all their recoverable natural gas.

Depropanization   Distillation in which lighter components are separated from butanes and higher boiling material, also called depropanation.

Desorption   The reverse process of adsorption whereby adsorbed matter is removed from the adsorbent, also used as the reverse of absorption (*q.v.*).

Developed proved reserves   Reserves that are expected to be recovered in existing wells, including reserves behind pipe, which may be recovered with the current infrastructure through additional work and with moderate investment costs. Reserves associated with secondary and/or enhanced recovery processes will be considered as developed when the infrastructure required for the process has been installed or when the costs required for such are lower. This category includes reserves in completed intervals that had been opened at the time when the estimation was made but that have not started flowing due to market conditions, connection problems, or mechanical problems and whose rehabilitation cost is relatively low.

Development   Activity that increases or decreases reserves by means of drilling exploitation wells.

Development well   A well drilled in a proved area in order to produce hydrocarbons.

Dew point   The temperature below which the water vapor in a volume of humid gas at a given constant barometric pressure will condense into liquid water at the same rate at which it evaporates.

Directional drilling   The technique of drilling at an angle from a surface location to reach a target formation not located directly underneath the well pad.

Discovered resource   The volume of hydrocarbons tested through wells drilled.

Discovery   The incorporation of reserves attributable to drilling exploratory wells that test hydrocarbon-producing formations.

Disposal well   A well which injects produced water into a regulated and approved deep underground formation for disposal.

Drilling mud   A fluid used to aid the drilling of boreholes; a mixture of clay, water, and other ingredients that are pumped downhole through the drill pipe and drill bit that enable the removal of the drill cuttings from the wellbore and also stabilize the penetrated rock formations before casing is installed in the borehole.

Drilling rig   The machine that creates the holes in the ground—typically large standing structures.

Dry gas   Natural gas containing negligible amounts of hydrocarbons heavier than methane. Dry gas is also obtained from the processing complexes.

Economic limit   The point at which the revenues obtained from the sale of hydrocarbons match the costs incurred in their exploitation.

Economic reserves   The accumulated production that is obtained from a production forecast in which economic criteria are applied.

Effective permeability   A relative measure of the conductivity of a porous medium for a fluid when the medium is saturated with more than one fluid. This implies that the effective permeability is a property associated with each reservoir flow, e.g., gas, oil, and water. A fundamental principle is that the total of the effective permeability is less than or equal to the absolute permeability.

Effective porosity   A fraction that is obtained by dividing the total volume of communicated pores by the total rock volume.

Emission   Air pollution discharge into the atmosphere, usually specified by mass per unit time.

Emission control   The use of gas cleaning processes to reduce emissions.

Emission standard   The maximum amount of a specific pollutant permitted to be discharged from a particular source in a given environment.

End-of-pipe emission control   The use of specific emission control processes to clean gases after production of those gases.

EPA   Environmental Protection Agency.

EPACT (Energy Policy Act of 1992)   Comprehensive energy legislation designed to expand natural gas use by allowing wholesale electric transmission access and providing incentives to developers of clean fuel vehicles.

EPCRA   Emergency Planning and Community Right-to-Know Act

Estimated additional amount in place   The volume additional to the proved amount in place that is of foreseeable economic interest. Speculative amounts are not included.

Estimated additional reserves recoverable   The volume within the estimated additional amount in place which geological and engineering information indicates with reasonable certainty might be recovered in the future.

Exploration   The process of identifying a potential subsurface geologic target formation and the active drilling of a borehole designed to assess the natural gas or oil.

Exploratory well   A well that is drilled without detailed knowledge of the underlying rock structure in order to find hydrocarbons whose exploitation is economically profitable.

Fabric filters   Filters made from fabric materials and used for removing particulate matter from gas streams (see also Baghouse).

Facies   One or more layers of rock that differ from other layers in composition, age, or content.

Fault   A fractured surface of geological strata along which there has been differential movement; a fracture surface in rocks along which movement of rock on one side has occurred relative to rock on the other side.

Fire point   The temperature to which gas must be heated under prescribed conditions of the method to burn continuously when the mixture of vapor and air is ignited by a specified flame.

Flash point   The temperature to which gas must be heated under specified conditions to give off sufficient vapor to form a mixture with air that can be ignited momentarily by a specified flame, dependent on the composition of the gas and the presence of other hydrocarbon constituents.

Flow-back (flowback) water   The water that returns to the surface from the wellbore within the first few weeks after hydraulic fracturing. It is composed of fracturing fluids, sand, and water from the formation, which may contain hydrocarbons, salts, minerals, or naturally occurring radioactive materials.

Flow line   A small-diameter pipeline that generally connects a well to the initial processing facility.

Formation (geologic)   A rock body distinguishable from other rock bodies and useful for mapping or description; formations may be combined into groups or subdivided into members.

Fracking   The process of injecting high pressured fluid containing water, sand, and chemicals into subsurface rock formations—the fluid fractures the rocks, improving the flow of natural gas into the wellbore.

Frac tank   The container in which the water or the proppant is held while a well is being fractured.

Fracture   A natural or man-made crack in a reservoir rock.

Fracturing   The breaking apart of reservoir rock by applying very high fluid pressure at the rock face; see also Fracking.

Free associated gas   Natural gas that overlies and is in contact with the crude oil of the reservoir—it may be gas cap.

Friction reducer   An additive that reduces the friction of a fluid as it flows through small spaces.

Gas condensate   See Condensate.

Gaseous pollutants   Gases released into the atmosphere that act as primary or secondary pollutants.

Gas hydrate   A molecule consisting of an ice lattice or cage in which low molecular weight hydrocarbon molecules like methane are embedded.

Gas-in-place (GIP)   The hypothetical amount of gas contained in a formation or rock unit; it always represents a value that is more than what is economically recoverable and refers to the total resources that are possible.

Gas processing   The preparation of gas for consumer use by removal of the non-methane constituents, synonymous with gas refining.

Gas refining   See Gas processing.

Geological province   A region of large dimensions characterized by similar geological history and development history.

Geological survey   The exploration for natural gas that involves a geological examination of the surface structure of the earth to determine the areas where there is a high probability that a reservoir exists.

Geophones   Equipment used to detect the reflection of seismic waves during a seismic survey.

Global warming   An environmental issue that deals with the potential for global climate change due to increased levels of atmospheric greenhouse gases.

Glycol-amine gas treating   A continuous, regenerative process to simultaneously dehydrate and remove acid gases from natural gas or refinery gas.

Greenhouse effect   Warming of the earth due to entrapment of the sun's energy by the atmosphere. See also Global warming.

Greenhouse gases   Gases that contribute to the greenhouse effect (*q.v.*). See also Global warming.

Groundwater   Water located beneath the surface of the earth: subsurface water that is in the zone of saturation and is the source of water for wells, seepage, and springs; the top surface of the groundwater is the *water table.*

GWPC   Ground Water Protection Council.

HAP(s)   Hazardous air pollutant(s).

HCPV   Hydrocarbon pore volume.

Heat of combustion (energy content)   The amount of energy that is obtained from burning natural gas, measured in Btu.

Heat value   The amount of heat released per unit of mass or per unit of volume when a substance is completely burned. The heat power of solid and liquid fuels is expressed in calories per gram or in Btu per pound. For gases, this parameter is generally expressed in kilocalories per cubic meter or in Btu per cubic foot.

Heterogeneity   In the current context, lack of uniformity in reservoir properties such as permeability.

HHV (*gross energy value, upper heating value, gross calorific value, higher calorific value*)   The same value as the thermodynamic heat of combustion since the enthalpy change for the reaction assumes a common temperature of the compounds before and after combustion, in which case the water produced by combustion is liquid.

Horizontal drilling   A drilling procedure in which the wellbore is drilled vertically to a kick-off depth above the target formation and then angled through a wide 90° arc such that the producing portion of the well extends horizontally through the target formation.

Horsehead (balanced conventional beam, sucker rod) pump   A common type of cable rod lifting equipment for recovery of oil and gas, so-called because of the shape of the counter weight at the end of the beam.

Hydraulic fracturing (fracking, fracing, fraccing)   A stimulation technique performed on low-permeability reservoirs, such as shale to increase oil and/or gas flow from the formation and improve productivity. Fluids and proppant are (*q.v.*) injected at high pressure and flow rate into a reservoir to create fractures perpendicular to the wellbore according to the natural stresses of the formation and maintain those openings during production.

Hydrocarbon   An organic compound containing only carbon and hydrogen. Hydrocarbons often occur in petroleum products, natural gas, and coals.

Hydrocarbon compounds   Chemical compounds containing only carbon and hydrogen.

Hydrocarbon resource   Resources, such as petroleum and natural gas, which can produce naturally occurring hydrocarbons without the application of conversion processes.

Hydrology   The study of water.

Hydrostatic pressure   The pressure exerted by a fluid at rest due to its inherent physical properties and the amount of pressure being exerted on it from outside forces.

Ideal gas   A gas in which all collisions between atoms or molecules are perfectly elastic and in which there are no intermolecular attractive forces.

Impure natural gas   Natural gas as delivered from the well and before processing (refining).

Independent producer   A nonintegrated company which receives nearly all of its revenues from production at the wellhead; by the IRS definition, a firm is an independent if the refining capacity is <50,000 barrels per day in any given day or their retail sales are <$5 million for the year.

Intermediate casing   Casing used on longer drilling intervals—set after the surface casing and before the production casing and prevents caving of weak or abnormally pressured formations.

IOGCC   Interstate Oil and Gas Commission.

Isopach   A line on a map designating points of equal formation thickness.

Kerogen   A complex carbonaceous (organic) material that occurs in sedimentary rock and shale, generally insoluble in common organic solvents.

Kitchen   The underground deposit of organic debris that is eventually converted to petroleum and natural gas.

Kriging   A technique used in reservoir description for interpolation of reservoir parameters between wells based on random field theory.

Lean gas   Natural gas in which methane is the major constituent.

Lease   A legal document that conveys to an operator the right to drill for oil and gas. Also, the tract of land on which a lease has been obtained and where producing wells and production equipment are located.

Liquefied natural gas   The liquid form of natural gas.

Liquefied petroleum gas (LPG)   Hydrocarbons, primarily composed of propane and butane, obtained during processing of crude oil, which are liquefied at low temperatures and moderate pressure. It is similar to natural gas liquid but originates from crude oil sources.

Lithology   The geological characteristics of the reservoir rock, the study of rocks. It is important for exploration and drilling crews to have an understanding of lithology as it relates to the production of gas and oil.

Logging   Lowering of different types of measuring instruments into the wellbore and gathering and recording data on *porosity*, *permeability*, and types of fluids present near the current well after which the data are used to construct subsurface maps of a region to aid in further exploration.

MACT (maximum achievable control technology)   This applies to major sources of hazardous air pollutants.

Magnetometer   A device to measure small changes in the Earth's magnetic field at the surface, which indicates what kind of rock formations might be present underground.

Marcellus Shale   A rock formation that extends from the base of the Catskills in New York and extends southwest to West Virginia, Kentucky, and Ohio.

Mcf (thousand cubic feet)   A unit of measure that is more commonly used in the low volume sectors of the gas industry.

MCL   Maximum contaminant level as dictated by regulations.

Membrane technology   Gas separation processes utilizing membranes that permit different components of a gas to diffuse through the membrane at significantly different rates.

MER   See Most efficient recovery rate.

Metamorphic rocks   Rocks resulting from the transformation that commonly takes place at great depths due to pressure and temperature. The original rocks may be sedimentary, igneous, or metamorphic.

Methane (CH_4)   Commonly (often incorrectly) known as natural gas, colorless and naturally odorless, and burns efficiently without many byproducts.

Methanogens   Methane-producing microorganisms.

Microseismic   The process of using seismic recording devices to measure the location of fractures that are created during the hydraulic fracing process; mapping of these microseismic events allows the extent of fracture development to be determined.

Migration (primary)   The movement of hydrocarbons (oil and natural gas) from mature, organic-rich source rocks to a point where the oil and gas can collect as droplets or as a continuous phase of liquid hydrocarbon.

Migration (secondary)   The movement of the hydrocarbons as a single, continuous fluid phase through water-saturated rocks, fractures, or faults followed by accumulation of the oil and gas in sediments (traps, *q.v.*) from which further migration is prevented.

Million 1×10^6.

Mineral rights   The rights of the owner of the property to mine or produce any resources below the surface of the property.

Most efficient recovery rate (MER)   The rate at which the greatest amount of natural gas may be extracted without harming the formation itself.

MSDS   Material Safety Data Sheet.

Muds   Used in drilling to lubricate the drilling bit in rotary drilling rigs.

Multilateral drilling   A drilling technique that is similar to stacked drilling in that it involves the drilling of two or more horizontal wells from the same vertical wellbore and the horizontal wells access different areas of the shale at the same depth, but in different directions.

Multiple completions   The result of drilling several different depths from a single well to increase the rate of production or the amount of recoverable gas.

NAAQS   National Ambient Air Quality Standards. Such standards exist for the pollutants known as the criteria air pollutants: nitrogen oxides (NO_x), sulfur oxides (SO_x), lead, ozone, particulate matter <10 microns in diameter, and carbon monoxide (CO).

Natural gas   A naturally occurring gas mixture consisting of methane and other hydrocarbons, used as an energy source to heat buildings, generate electricity and recently, to power motor vehicles.

Natural gas liquids (NGL)   Hydrocarbons, typically composed of propane, butane, pentane, hexane, and heptane, obtained from natural gas production or processing which are liquefied at low temperatures and moderate pressure. They are similar to LPG but originate from natural gas sources.

Natural Gas Act   Law passed in 1938 giving the Federal Power Commission (now the Federal Energy Regulatory Commission or FERC) jurisdiction over companies engaged in the interstate sale or transportation of natural gas.

Natural gasoline   A mixture of liquid hydrocarbons extracted from natural gas (*q.v.*) suitable for blending with refinery gasoline.

Natural Gas Policy Act of 1978   One of the first efforts to deregulate the gas industry and to determine the price of natural gas as dictated by market forces, rather than regulation.

Natural Gas Resource Base   An estimate of the amount of natural gas available, based on the combination of proved reserves, and those additional volumes that have not yet been discovered, but are estimated to be "discoverable" given current technology and economics.

NES (National Energy Strategy)   A 1991 federal proposal that focused on national security, conservation, and regulatory reform, with options that encourage natural gas use.

NESHAP (National Emissions Standards for Hazardous Air Pollutants)   Emission standards for specific source categories that emit or have the potential to emit one or more hazardous air pollutants; the standards are modeled on the best practices and most effective emission reduction methodologies in use at the affected facilities.

Net thickness   The thickness resulting from subtracting the portions of the reservoir that have no possibilities of producing hydrocarbon from the total thickness.

Nonassociated gas   Natural gas found in reservoirs that do not contain crude oil at the original pressure and temperature conditions, sometimes called *gas well gas*, gas produced from geological formations that typically do not contain much, if any, crude oil, or higher boiling hydrocarbons (*gas liquids*) other than methane; can contain nonhydrocarbon gases, such as carbon dioxide and hydrogen sulfide.

Nonproved reserves   Volumes of hydrocarbons and associated substances, evaluated at atmospheric conditions, resulting from the extrapolation of the characteristics and parameters of the reservoir beyond the limits of reasonable certainty or from the assumption of oil and gas forecasts with technical and economic scenarios other than those in operation or with a project in view.

Normal fault   The result of the downward displacement of one of the strata from the horizontal. The angle is generally between 25 and 60° and it is recognized by the absence of part of the stratigraphic column.

Observation wells   Wells that are completed and equipped to enable measurement of reservoir conditions and/or sample reservoir fluids, rather than for injection or production of reservoir fluids.

Oil shale   A fine-grained impervious sedimentary rock which contains an organic material called kerogen.

Olamine process   A process that uses an amine derivative (an olamine) to remove acid gas from natural gas streams.

Olamines   Compounds such as ethanolamine (monoethanolamine, MEA), diethanolamine (DEA), triethanolamine (TEA), methyldiethanolamine (MDEA), diisopropanolamine (DIPA), and diglycolamine (DGA) which are widely used in gas processing.

Organic sedimentary rocks   Rocks containing organic materials, such as residues of plant and animal remains/decay.

Original gas volume in place   The amount of gas that is estimated to exist initially in the reservoir and that is confined by geologic and fluid boundaries, which may be expressed at reservoir or atmospheric conditions.

Original pressure   The pressure prevailing in a reservoir that has never been produced. It is the pressure measured by a discovery well in a producing structure.

Original reserve   The volume of hydrocarbons at atmospheric conditions that are expected to be recovered economically by using the exploitation methods and systems applicable at a specific date. It is a fraction of the discovered and economic reserve that may be obtained at the end of the reservoir exploitation.

Pad drilling   A technique in which a drilling company uses a single drill pad to develop as large an area of the subsurface as possible.

Particulate matter (particulates)   Particles in the atmosphere or on a gas stream that may be organic or inorganic and originate from a wide variety of sources and processes.

Perforation   A hole in the casing, often generated by means of explosive charges, which enables fluid and gas flow between the wellbore and the reservoir.

Permeability   A measure of the ability of a material to allow fluids to pass through it; it is dependent upon the size and shape of pores and interconnecting pore throats; a rock may have significant porosity (many microscopic pores) but have low permeability if the pores are not interconnected; permeability may also exist or be enhanced through fractures that connect the pores.

Petroleum (crude oil)   A naturally occurring mixture of gaseous, liquid, and solid hydrocarbon compounds usually found trapped deep underground beneath impermeable cap rock and above a lower dome of sedimentary rock like shale; most petroleum reservoirs occur in sedimentary rocks of marine, deltaic, or estuarine origin.

Physical limit   The limit of the reservoir defined by any geological structures (faults, unconformities, change of facies, crests, and bases of formations, etc.), caused by contact between fluids or by the reduction to critical porosity of permeability limits or by the compound effect of these parameters.

Play   A group of fields sharing geological similarities where the reservoir and the trap control the distribution of oil and gas; a geologic area where hydrocarbon accumulations occur—also called a *resource (q.v.)*; for shale gas, examples include the Barnett and Marcellus plays.

Pollutant   A chemical (or chemicals) introduced into land, water, and air systems that is (are) not indigenous to these systems, also an indigenous chemical (or chemicals) introduced into land, water, and air systems in amounts greater than the natural abundance.

Pollution   The introduction into land, water, and air systems of a chemical or chemicals that are not indigenous to these systems or the introduction into land, water, and air systems of indigenous chemicals in greater-than-natural amounts.

Pooling or unitization   A provision that allows landowners to combine land to form a drilling unit.

Pore space   A small hole in reservoir rock that contains fluid or fluids; a 4-inch cube of reservoir rock may contain millions of interconnected pore spaces.

Pore volume   The total volume of all pores and fractures in a reservoir or part of a reservoir.

Porosity   The percentage of void space in a rock that may or may not contain oil or gas.

Possible reserves   Reserves where there is an even greater degree of uncertainty but about which there is some information.

Potential reserves   Reserves based upon geological information about the types of sediments where such resources are likely to occur and they are considered to represent an educated guess.

Pressure cores   Cores cut into a special coring barrel that maintains reservoir pressure when brought to the surface; this prevents the loss of reservoir fluids that usually accompanies a drop in pressure from reservoir to atmospheric conditions.

Primary term   The length of a lease in years.

Probable reserves   Mineral reserves mineral that are nearly certain but about which a slight doubt exists.

Produced water   The water that is brought to the surface during the production of oil and gas. It typically consists of water already existing in the formation, but may be mixed with fracturing fluid if hydraulic fracturing was used to stimulate the well.

Producer   The company generally involved in exploration, drilling, and refining of natural gas.

Producibility   The rate at which oil or gas can produced from a reservoir through a wellbore.

Producing well   A well in an oil field or gas field used for removing fluids from a reservoir.

Production casing   The final interval in a well and the smallest casing which forms the outer boundary of the annulus.

Production rate   The rate of production of oil and/or gas from a well, usually given in barrels per day (bbls/day) for oil or standard cubic feet (scft^3/day) for gas.

Proppant/propping agent   Particles mixed with fracturing fluid to maintain fracture openings after hydraulic fracturing; these typically include sand grains, but they may also include engineered proppants; silica sand or other particles pumped into a formation during a hydraulic fracturing operation to keep fractures open and retain the induced permeability.

Prospective resource   The amount of hydrocarbons evaluated at a given date of accumulations not yet discovered but which have been inferred and are estimated as recoverable. See also Undiscovered resource.

Proved area   The known part of the reservoir corresponding to the proved volume.

Proved reserves (proven reserves)   Mineral reserves that have been positively identified as recoverable with current technology.

Proved resources   Part of the resource base that includes the working inventory of natural gas; volumes that have already been discovered and are readily available for production and delivery.

Proved amount in place   The volume originally occurring in known natural reservoirs which has been carefully measured and assessed as exploitable under present and expected local economic conditions with existing available technology.

Proved recoverable reserves   The volume within the proved amount in place that can be recovered in the future under present and expected local economic conditions with existing available technology.

Quad   An abbreviation for a quadrillion (1,000,000,000,000,000) Btu; roughly equivalent to 1 trillion (1,000,000,000,000) cubic feet, or 1 Tcf. See also, Bcf, Mcf, Tcf.

Quadrillion 1×10^{15}.

RACT (Reasonably Available Control Technology standard)   Implemented in areas of nonattainment to reduce emissions of volatile organic compounds and nitrogen oxides.

Raw natural gas   Impure natural gas as delivered from the well and before processing (refining).

Recovery factor   The ratio between the original volume of oil or gas at atmospheric conditions and the original reserves of the reservoir.

Reduced emission completion (REC or green completion)   An alternative practice that captures and separates natural gas during well completion and workover activities instead of allowing it to vent into the atmosphere.

Relative permeability   The permeability of rock to gas, oil, or water, when any two or more are present, expressed as a fraction of the air phase permeability of the rock.

Remaining reserves   The volume of hydrocarbons measured at atmospheric conditions that are still to be commercially recoverable from a reservoir at a given date, using the applicable exploitation techniques. It is the difference between the original reserve and the cumulative hydrocarbon production at a given date.

Reserve additions   Volumes of the resource base that are continuously moved from the resource category to the proved resources category.

Reserve replacement rate   A rate that indicates the amount of hydrocarbons replaced or incorporated by new discoveries compared with what has been produced in a given period. It is the coefficient that arises from dividing the new discoveries by production during the period of analysis. It is generally referred to in annual terms and is expressed as a percentage.

Reserve–production ratio   The result of dividing the remaining reserve at a given date by the production in a period. This indicator assumes constant production, hydrocarbon prices, and extraction costs, without variation over time, in addition to the nonexistence of new discoveries in the future.

Reserves   Well-identified resources that can be profitably extracted and utilized with existing technology, the estimated volume of gas economically recoverable from single or multiple reservoirs. Reserve estimates are based on strict site-specific engineering criteria.

Reservoir   An area that contains a resource. In fracking, well operators are seeking to tap into natural gas reservoirs deep underground.

Reservoir energy   The underground pressure in a reservoir that will push the petroleum and natural gas up the wellbore to the surface.

Reservoir simulation   Analysis and prediction of reservoir performance with a computer model.

Residue gas   Natural gas from which the higher molecular weight hydrocarbons have been extracted, mostly methane.

Resource   The total amount of a commodity (usually a mineral but can include nonminerals, such as water and petroleum) that has been estimated to be ultimately available, also called a *play (q.v.)*.

Reverse fault   The result of compression forces where one of the strata is displaced upward from the horizontal.

Revision   The reserve resulting from comparing the previous year's evaluation with the new one, in which new geological, geophysical, operation, and reservoir performance information is considered, in addition to variations in hydrocarbon prices and extraction costs. It does not include well drilling.

Rich gas   A gaseous stream traditionally very rich in natural gas liquids; see also Natural gas liquids.

Rock matrix   The granular structure of a rock or porous medium.

Royalty   A payment received by the lessor from the oil or gas company, based on the production of the well and market prices.

R/P (reserves/production) ratio   Calculated by dividing proved recoverable reserves by production (gross less reinjected) in a given year.

Sand   A course granular mineral mainly comprising quartz grains that is derived from the chemical and physical weathering of rocks rich in quartz, notably sandstone and granite.

Sandstone   A sedimentary rock formed by compaction and cementation of sand grains; can be classified according to the mineral composition of the sand and cement.

Scrubbing   Purifying a gas by washing with water or chemical; also but less frequently used to describe the removal of entrained materials.

Secondary pollutants   A pollutant (chemical species) produced by interaction of a primary pollutant with another chemical or by dissociation of a primary pollutant or by other effects within a particular ecosystem.

Secondary term   The length of a lease after a well is drilled.

Sedimentary   Formed by or from deposits of sediments, especially from sand grains or silts transported from their source and deposited in water, as sandstone and shale, or from calcareous remains of organisms, as limestone.

Sedimentary strata   Typically consist of mixtures of clay, silt, sand, organic matter, and various minerals; formed by or from deposits of sediments, especially from sand grains or silts transported from their source and deposited in water, such as sandstone and shale; or from calcareous remains of organisms like limestone.

Seismic event   An earthquake-induced seismicity is an earthquake caused by human activities.

Seismic section   A seismic profile that uses the reflection of seismic waves to determine the geological subsurface.

Seismograph   An instrument used to detect and record earthquakes that is able to pick up and record the vibrations of the earth that occur during an earthquake; when seismology is applied to the search for natural gas, seismic waves, emitted from a source, are sent into the earth and the seismic waves interact differently with the underground formation (underground layers), each with its own properties.

Seismology   The study of the movement of energy, in the form of seismic waves, through the Earth's crust.

Shale A fine-grained sedimentary rock that is formed from compacted mud—black shale sometimes breaks down to form natural gas or oil.

Shale gas Natural gas stored in low-permeability shale formations; see also Unconventional gas.

Shut-in royalty A payment to the lessor in lieu of a production royalty. This is received when a well cannot produce due to production problems.

Sonic log A well log based on the time required for sound to travel through rock, useful in determining porosity.

Sour gas Natural gas that contains hydrogen sulfide.

Spacing The optimum distance between hydrocarbon-producing wells in a field or reservoir.

Stacked wells Drilling at the horizontal where shale is sufficiently thick or multiple shale rock strata are found layered on top of each other; one vertical wellbore can be used to produce gas from horizontal wells at different depths.

Standard conditions The reference amounts for pressure and temperature—in the British system, they are 14.73 pounds per square inch for the pressure and 60°F for temperature.

Stimulation Any of several processes used to enhance near reservoir permeability.

Strata Layers including the solid iron-rich inner core, molten outer core, mantle, and crust of the earth.

Stratigraphy The subdiscipline of geology that studies the origin, composition, distribution, and succession of rock strata.

Stripper wells Natural gas wells that produce <60,000 cubic feet of gas per day.

Surface casing A pipe that protects freshwater aquifers and also provides structural strength so that other casings may be used.

Surfactant A compound that lowers the surface tension of a liquid.

Sweetening process A process for the removal of hydrogen sulfide and other sulfur compounds from natural gas.

Sweet gas Natural gas that contains very little, if any, hydrogen sulfide.

Tcf (trillion cubic feet) Gas measurement approximately equal to 1 quadrillion (1,000,000,000,000,000) Btu.

Technical reserves The accumulative production derived from a production forecast in which economic criteria are not applied.

Termination The end of a lease.

Thermogenic gas Gas formed by pressure effects and temperature effects on organic debris.

Three-Dimensional (3-D) Seismic Survey This allows producers to see into the Earth's crust to find promising formations for retrieval of gas.

Time-lapse logging The repeated use of calibrated well logs to quantitatively observe changes in measurable reservoir properties over time.

Total thickness The thickness from the top of the formation of interest down to a vertical boundary determined by a water level or by a change of formation.

Tracer test A technique for determining fluid flow paths in a reservoir by adding small quantities of easily detected material (often radioactive) to the flowing fluid and monitoring their appearance at production wells.

Transgression A geological term used to define the immersion of one part of the continent under sea level as a result of a descent of the continent or an elevation of the sea level.

Transmissibility (transmissivity) An index of producibility of a reservoir or zone, the product of permeability and layer thickness.

Traps A generic term for an area of the Earth's crust that has developed in such a way as to *trap* gas beneath the surface.

TRI Toxics release inventory.

Triaxial borehole seismic survey A technique for detecting the orientation of hydraulically induced fractures, wherein a tool holding three mutually seismic detectors is clamped in the borehole during fracturing; fracture orientation is deduced through analysis of the detected microseismic perpendicular events that are generated by the fracturing process.

Trillion 1×10^{12}.

Trillion cubic feet A volume measurement of natural gas, approximately equivalent to 1 quad.

Ultimate recovery The cumulative quantity of oil and/or that will be recovered when revenues from further production no longer justify the costs of the additional production.

Unconformity A surface of erosion that separates younger strata from older rocks.

Unconventional gas Gas that occurs in tight sandstones, siltstones, sandy carbonates, limestone, dolomite, and chalk; see also Shale gas.

Undeveloped proved area   The plant projection of the extension drained by the future producing wells of a producing reservoir and located within the undeveloped proved reserve.

Undeveloped proved reserves   The volume of hydrocarbons that is expected to be recovered through wells without current facilities for production or transportation and future wells. This category may include the estimated reserve of enhanced recovery projects, with pilot testing or with the recovery mechanism proposed in operation that has been predicted with a high degree of certainty in reservoirs that benefit from this kind of exploitation.

Undiscovered resource   The volume of hydrocarbons with uncertainty but whose existence is inferred in geological basins through favorable factors resulting from the geological, geophysical, and geochemical interpretation. They are known as prospective resources (*q.v.*) when considered commercially recoverable.

Unsaturated zone   A zone where the soil and the rock contains air as well as water in its pores and which is above the groundwater table. The unsaturated zone does not contain readily available water, but it does provide water and nutrients to the biosphere.

Utica Shale   A natural gas containing rock formation below the Marcellus Shale. The Utica Shale formation extends from eastern Ohio through much of Pennsylvania to western New York.

Vadose zone   The layer of earth between the land's surface and the position of groundwater at atmospheric pressure.

Vapor density   The density of any gas compared to the density of air with the density of air equal to unity.

Vertical seismic profiling   A method of conducting seismic surveys in the borehole for detailed subsurface information.

Viscosity   The measure of a fluid's thickness or how well it flows.

VOCs (volatile organic compounds)   Compounds regulated because they are precursors of ozone; carbon-containing gases and vapors from incomplete gasoline combustion and from the evaporation of solvents.

VSP (vertical seismic profiling)   A method of conducting seismic surveys in the borehole for detailed subsurface information.

Well abandonment   The final activity in the operation of a well when it is permanently closed under safety and environment preservation conditions.

Wellbore (well bore)   The hole in the earth comprising a well—this includes the inside diameter of the drilled hole bounded by the rock face.

Well casing   A series of metal tubes installed in the freshly drilled hole serving to strengthen the sides of the well hole, ensuring that no oil or natural gas seeps out of the well hole as it is brought to the surface, and keeping other fluids or gases from seeping into the formation through the well.

Well completion   The process for completion of a well to allow for the flow of petroleum or natural gas out of the formation and up to the surface; includes strengthening the well hole with casing, evaluating the pressure and temperature of the formation, and then installing the proper equipment to ensure an efficient flow of natural gas out of the well; the complete outfitting of an oil well for either oil production or fluid injection; also the technique used to control fluid communication with the reservoir.

Well head (wellhead)   The pieces of equipment mounted at the opening of the well to regulate and monitor the extraction of hydrocarbons from the underground formation; prevents leaking of oil or natural gas out of the well and prevents blowouts due to high pressure formations; the structure on the well at ground level that provides a means for installing and hanging casing, production tubing, flow control equipment, and other equipment for production.

Well logging   A method used for recording rock and fluid properties to find gas and oil containing zones in subterranean formations.

Well logs   The information concerning subsurface formations obtained by means of electric, acoustic, and radioactive tools inserted in the wells. The logs also include information about drilling and the analysis of mud and cuts, cores, and formation tests.

Wet gas   Gas containing a relatively high proportion of hydrocarbons which are recoverable as liquids; see also Lean gas.

Wet scrubbers   Devices in which a counter-current spray liquid is used to remove impurities and particulate matter from a gas stream.

Wobbe number (Wobbe Index)   The calorific value of a gas divided by the specific gravity.

Workover   The repair or refracturing of an existing oil or gas well to enhance or prolong production.

Lightning Source UK Ltd.
Milton Keynes UK
UKOW04f2113250414

^586UK00019B/237/P